T0290650

Syria, 12th or 13th century A.D.
Gold sheet, engraved, appliqués, almandine garnet
The Nasser D. Khalili Collection of Islamic Art
Accession no. JLY206

Turquoise Mamluk ring
Courtesy of Christie's Images, London

ARAB ROOTS OF GEMOLOGY

Ahmad ibn Yusuf Al Tifaschi's
Best Thoughts on the Best of Stones

Translated and with commentary by
Samar Najm Abul Huda

The Scarecrow Press, Inc.
Lanham, Md., & London
1998

SCARECROW PRESS, INC.

Published in the United States of America
by Scarecrow Press, Inc.
4720 Boston Way
Lanham, Maryland 20706

4 Pleydell Gardens, Folkestone
Kent CT20 2DN, England

British Library Cataloguing in Publication Information Available

Library of Congress Cataloging-in-Publication Data

Tīfāshī, Ahmad ibn Yusuf, d. 1253 or 4.
[Azhār al-afkār fī jawāhir al-ahjār. English]
Arab roots of gemology : Ahmad ibn Yusuf al Tifaschi's Best thoughts on the best of stones / translated and with commentary by Samar Najm Abul Huda.
p. cm.
Includes bibliographical references and index.
ISBN 0-8108-3294-1 (cloth : alk. paper)
1. Precious stones—Early works to 1800. 2. Gem cutting—Arab countries—Early works to 1800. I. Title.
QE392.T513 1998
553.8—dc21 97-9085

ISBN 0–8108–3294–1 (cloth : alk. paper)

The paper used in this publication meets the minimum requirements of American National Standard for Information Sciences—Permanence of Paper for Printed Library Materials, ANSI Z39.48–1984.
Manufactured in the United States of America.

IF THERE WERE TWO STONES MORE PRECIOUS THAN DIAMOND AND RUBY, I WOULD CALL THEM TAMARA AND NADINE. TO MY DAUGHTERS WITH ALL MY LOVE.

S.A.H.

Contents

Foreword

ALTHOUGH I was aware that well before the thirteenth century, the Arabs were making superb scientific instruments–astrolables–for measuring star altitudes and time, I was nevertheless unprepared to read about how far they had developed scientific gemology in that century, as disclosed by the publication of research by Samar Najm Abul Huda.

One man in particular, Ahmad ibnYusuf al Tifaschi, born in 1184, a judge, scholar, and poet, was first taught about gemstones by his father. Thus inspired, he began reading all the earlier writings on the subject he could discover, including those of Aristotle, Theophrastus, and Pliny. He also traveled extensively to the gem mining and trading centers of his time and made contact with various European gem experts and merchants. About ten years before his death in 1253, he wrote what was later described as the most "methodical and complete" work on precious stones. It contained information that seems surprisingly advanced even to the gemologist of today. As Abul Huda points out, al Tifaschi's account was based on his own observations and experiments and was not a restatement of what Pliny and others had written. As was the custom of the time, however, he also included supposed medicinal properties and superstitious beliefs attached to the stones.

The book covers twenty-five different stones, giving the crystallographic forms and physical properties such as specific gravity, hardness, luster, dispersion, including habit, and inclusions. He even differentiated false from true cleavage and described pleochroism, as well as ways of determining genuine and imitation stones. The book describes causes of the formation of gemstones within their mother rocks as well as characteristics of the mine concerned. Even the marketplace is not forgotten: He provided average prices of the stones considered.

Abul Huda's study and translation of al Tifaschi's fascinating work is an important contribution to our knowledge of the history of gemology, illustrating that as a scientific discipline it has existed for over seven centuries, not hardly one, as most modern gemologists probably believe.

Eric Bruton

Former President, Gemological Association of Great Britain

Foreword

Preface

I decided to write a book on gemology in 1989. At that time, I was taking a course in gemology in Athens, Greece, and my father had sent me a book on this subject written by a relatively unknown thirteeth-century Arab scholar. This book left a deep impression on my mind, as this Arab scholar had advanced theories and concepts as early as the year A.H. 640 (A.D. 1240) that laid the early groundwork upon which several modern concepts of gemology are based. Living in Greece since the early 1980s had consolidated my love for my Arab roots and heritage, as it often happens to people living away from their own country and culture, and, together with my thirst for knowledge and perhaps a writing urge inherited from my father, the idea of acquainting the West with the early Arab "gemologists" and their substantial contribution to that science began to take root. The idea hibernated in my mind for several years, and I began serious work on this book only in 1992.

The first book written by an Arab gemologist that came to my attention was *Azhar al Afkar fi Djawahir al Ahdjar* (Best Thoughts on the Best of Stones) by Ahmad ibn Yusuf al Tifaschi.

Al Tifaschi (A.D. 1184-1253) is little known both in the Arab world and in the West, despite the numerous books he left, the most important of which, in my opinion, is the above-mentioned one on gemology. A manuscript of this book is in the British Museum Library in London, having been presented to the museum in 1857 by a Major Perronet Thompson; eight other original copies of al Tifaschi's manuscript are in the Bibliothèque Nationale in Paris, France, although one only dates from the thirteeth century. An original manuscript of the above book is also in the Topkapi Saray in Istanbul, another one is in Cairo, and still another one is in Sheikh Khalil al Khalidy's library in Jerusalem.

The importance of this book was recognized in the West by some famous orientalists: Sebaldus Ravius of Utrecht, Holland, who first published the first three chapters of the book in Latin in 1784; Count Raineri Biscia of Italy, who first translated the book into Italian and published it in Bologna in 1818 under the title *Fior Di Pensi Sulle Pietre Preziosi Di Ahmed Teifascite;* Clément Mullet, who wrote an essay called "Essai sur la Minéralogie Arabe," where he translated part of al Tifaschi's book and published his paper in the *Journal Asiatique* in Paris in 1868; and finally by the German Julius Ruska, who wrote about al Tifaschi in 1912. The few Arab researchers who commented on al Tifaschi's book were J. al Haschmi in 1936; Father Anastase-Marie de Ste. Elie in 1939; and Drs. M. Y. Hassan and M. B. Khafaga, who wrote by far the best available study of al Tifaschi's book in 1977. However, none of the above authors was a gemologist, and the importance of many aspects of al Tifaschi's work has been so far unrecognized.

This book is different from these attempts to study al Tifaschi. For the first time, an Arab gemologist was studied by an Arab author; the research, however, was conducted in English. The obvious advantages with regard to the subtleties of language, nomenclature, and references will, it is hoped, be useful in acquainting the Western world with this important work.

This book is addressed to all readers interested in geology, gemology, mineralogy, jewelry, history, Arab heritage, Islamic art, and the history of science. The reader is neither supposed to have a scientific knowledge of gemology, nor be completely ignorant of the different gemstones available.

I have tried to remain as faithful as possible to the original text and as thorough as possible in my research. If, however, errors are still present, I ask the reader to forgive me, as this is a first effort in both writing and gemology.

I hope that this study will return al Tifaschi to his rightful place in the history of gemology, and that the name of Ahmad ibn Yusuf al Tifaschi will be as famous as that of many other Arab scholars whose contributions in the different fields have long been established.

This book is intended to be interesting, entertaining, and educational:

- Interesting because, at a time when Europe was painfully making progress in science, the East was enjoying the golden age of Islamic civilization, as evidenced by the profusion of writings in all kinds of subjects, including the one that interests us here, gemology.
- Entertaining because of the legends and lore attached to the stones, the various superstitions and talismanic and medicinal properties attributed to some of them (some of which are still believed in today), and the unorthodox methods used by the early gemologists for testing and improving the appearance of these stones.
- Educational because Ahmad al Tifaschi's book can be described as unknown in the West.

I hope the reader will enjoy reading this book as much as I enjoyed writing it.

S.A.H.

Acknowledgments

I wish to thank the many people without whose help this book would not have been possible.

The first person to have stirred my interest in gemology is Nicholas Kielty-Lambrinidis, former President of the Gemological Association of Greece, under whose tutorship I studied gemology in Athens. His kindness, patience and thorough knowledge of the subject were the catalyst for my idea of writing a book.

However the subject of this book would not have been the first Arab gemologists had it not been for my father, Dr. Mohammad Yusuf Najm, a distinguished Arab scholar and professor of Arabic literature at the American University of Beirut since 1954, who provided me with rare research and studies of manuscripts by the Arab gemologists of the ninth to fourteenth centuries, as well as all Arab references needed from his own collection of old and rare books on the subject.

Eric Bruton, formerly of N.A.G. Press, London, was the first to foresee the importance of some of the points unearthed by my research, and his encouragement throughout the various stages of completion of this book was unwavering. Allan E. Jobbins, former Curator, British Museum of Natural History, London, has reviewed the book and pointe ' out some errors for which I am deeply grateful. I also owe special thanks to Dr. John Sinkankas, of San Diego, who spent much time and effort relentlessly scrutinizing the final drafts of the manuscript to pinpoint any remaining discrepancies and offer relevant advice and expertise on the subject. This book would have definitely been incomplete without his help.

Finally, the King Faisal Center for Research and Islamic Studies in Saudi Arabia provided many documents related to the subject and was of great assistance in locating some rare manuscripts by the medieval Arab scholars.

I am deeply indebted to all these people for their valuable assistance in completing this book.

Samar Najm Abul Huda

Introduction

BY the early years of the eighth century A.D., the Muslim world, under the zealotry of Mohammad and his followers, stretched unbroken from western India through Persia, the Middle East, along the north coast of Africa, and terminated in the Iberian Peninsula. Expansion was essentially complete by the early years of the eighth century, and the time for consolidation of gains had come, along with the need to solve problems of the administration of peoples and lands of diverse character. However, according to Hitti (1969, 240), "the invaders from the desert brought with them no tradition of learning, no heritage of culture, to the lands they conquered. In Syria, in Egypt, in Iraq, in Persia, they sat as pupils at the feet of the peoples they subdued . . . and what acquisitive pupils they proved to be!"

The native drive that impelled former desert dwellers to emerge upon a world hitherto unknown to them now turned toward the acquisition and digestion of knowledge: "the seed was sown" for the growth of the "tree of knowledge," whose roots were embedded in the cultures of the Greeks, Syrians, and Persians, among others. Hitti (ibid., 363) now speaks of the Arab "epoch of translation," or the deliberate, sustained effort to acquire existing manuscripts in the languages of the conquered and translate them into Arabic, which, for the first time, became a major medium for the preservation and dissemination of ancient knowledge, cultural and scientific. This epoch spanned approximately the years A.D. 750-850, with major efforts devoted to translations from Persian, Sanskrit, Syriac, and Greek. This corresponds to the period that historians called "the preservation of knowledge" period, in allusion to the fact that many invaluable writings in other languages, destroyed in the vicissitudes of war and upheaval, were preserved in the Arabic language.

At this point, it should be observed that however beneficial such translations proved to be, thc barrier of language prevented their wider dissemination because of the utter foreignness of the Arabic alphabet to readers accustomed to the Roman alphabet. While readers in the Roman alphabet find many common word roots and other clues to meaning in other languages employing this alphabet, the fact is that the Arabic alphabet offers them nothing of the sort, and therefore Arabic remains an unfathomable mystery. If the Muslim world is to become better known through its preserved literature, it would appear highly desirable that a new "epoch of translation" take place, this time from Arabic to languages employing the Roman alphabet. The present translation of al Tifaschi is a much desired beginning.

One of the apprehensions that attends the writing of a bibliographical work is that some truly important book somehow has escaped the attention of the bibliographer. It is therefore with a sense of relief, coupled with gratitude, that al Tifaschi's splendid compilation of the thirteenth century is now before us, translated and annotated by a member of the Muslim world who is also expert in the subject that al Tifaschi discoursed upon so long ago! I express my relief because my remarks on al Tifaschi in my own gemological bibliography (1993, 224, 1030), dependent as they were upon translations by others, can now be augmented by Samar Abul Huda's definitive study. Mrs. Abul Huda has observed that all previous western studies of al Tifaschi were by students whose Arabic was acquired, not like hers, their native language. We therefore can read the present translation with the confidence that the meanings of its long-departed author come to us in purer form.

At the beginning of his treatise, al Tifaschi takes pains to assure his readers that nowhere else will they find such complete information on gemstones than in his book, stating that much of his information was derived from personal experience and experimentation and also from consultations with merchants and other experts whose opinions he valued. He also acknowledges his debt to ancient and contemporary writers from whom he has borrowed information, e.g., Pliny, *Natural History*, A.D.79. As an authority on gemstones, al Tifaschi's work was lauded by many and noted with favor by various modern historians, such as Hitti (383) and Sarton (vol. 2, 650).

In his survey of twenty-five substances, mostly from the mineral kingdom, al Tifaschi reflects their importance in his time by how much space he devotes to each. By far the most wordage is given to bezoars, those mysterious animal calculi taken from various glands and organs of animal bodies and thought to possess wonderful healing or preventive powers, especially against poisons. Although organic in origin, their value was deemed so high that they were customarily classed with their precious mineral counterparts. This large section ranks in importance with the monograph on bezoars written by Bauhin in 1613 (Sinkankas, 80). Another very large chapter, on pearls, is not unexpected in view of the centuries-old familiarity of the Muslims with the pearl oyster-producing waters of the Indian Ocean and the Persian Gulf. Another marine product, coral, is also accorded a large treatment. The mineral gemstones, notably corundum, emerald, and diamond, receive good treatment, but for a number of others the discussions are relatively brief. In the case of moonstone, described in chapter 9, al Tifaschi admits that "in any book on stones, whether old or new," he could find no mention of same, but his description led Abul Huda to conclude that the gemstone meant is moostone rather than the chrysoberyl cat's eye favored by Clément Mullet (1981; Sinkankas, 224).

In conclusion, al Tifaschi's treatise is a combination of scientific fact and liberal doses of gossip, speculation, and curious lore, and in many places it reminds one of the writing of Pliny in his *Natural History* but with less inclusion of the incredible. While some may object or ridicule the attention paid by early writers to such lore, it remains true that the beliefs in the curative or prophylactic value of gem materials are as much a part of our cultural heritage as those bits of knowledge that we assume to be fact. Al Tifaschi pleasantly provides both.

References:

HITTI, P. K. *History of the Arabs*. 7th ed. Macmillan, London, 1960.

SARTON, G. *Introduction to the History of Science*. Vol. 2. Carnegie Institution of Washington, D.C., 1947.

SINKANKAS, J. *Gemology: An Annotated Bibliography*. Scarecrow Press, Lanham, Md. 1993.

John Sinkankas
San Diego, California, 1996

PART I
GENERAL

CHAPTER 1

Background

THE first study of gemstones started probably well before 300 B.C.; however, the written proof of the interest in precious gems came to us through the work of Theophrastus around 300 B.C.

The famous Latin naturalist Pliny copied many of Theophrastus's ideas, and the latter's *Treatise on Stones* inspired Pliny's Book 37 of his *Historia Naturalis* dealing with gemstones and written around A.D. 70. After Pliny, and until the fifteenth century, the rare Western scholars who wrote on gemstones include the following: Plutarch of Chaeronea (A.D. 46-120), a Greek historian and philosopher, who summarized the knowledge of gemstones of his time; Isidore, Bishop of Seville (A.D. 570-636), who gathered extracts from Theophrastus and Pliny in his *Etymologia*; Michael Constantine Psellos (A.D. 1018-1076), a Byzantine historiographer; Albertus Magnus, Bishop of Regensburg (A.D. 1193-1280), who wrote about the healing magic of stones in his *De Mineralibus et Rebus Metalicis*; Alphonse de Castile born in the Lebanon (A.D. 1103-1148), who wrote *Lapidario*; and finally a person either called Aristotle or using that name, referred to as "Pseudo-Aristotle." This "Pseudo-Aristotle" was often quoted by all the Arab gemologists of the ninth to fourteenth centuries.

During this time, Arab scholars* were not idle, and many wrote extensively on gemstones, including such authors as 'Aoun al 'Ibadi, Ayub al Aswad al Basri, Ibn Shazan, Sabah and Ya'qub al Kindi, Nasr al Djawhari al Farissi, Ahmad ibn'Abdul'Aziz, 'Utared ibn Mohammad al Katib (or al Hasib?) and Hunayn ibn Ishaq al Hakim, (all during the Umayyad and 'Abbasid periods which lasted from the first to the seventh centuries A.H. or the seventh to thirteenth centuries A.D.).

The first scientific book written on gemstones was by Ya'qub al Kindi (A.D. 801-866 or 873?) who dealt with all aspects of gemology at that time

* By Arab scholars are meant here authors that are "Persians, Syrians, Egyptians, or Arabians, Christian, Jews, or Moslem, who may have drawn some of their material from Greek, Aramean, Indo-Persian or other sources," but who wrote books in the Arabic language.[1]

in his *Fil Djawahir wal Ashbah* (On Gemstones and their Imitations). He had also pointed out that, as many stones can be confused with each other, because they have similar colors or ranges of color, two methods can be used to establish the identity of a stone: Hardness (scratch test) and weight (specific gravity). Al Biruni followed in the eleventh century with *Al Djamahir fi Ma'arifat al Djawahir* (General Knowledge on Gemstones), a voluminous work of over 300 pages covering not only gemstones, but also minerals (gold, silver, copper, etc) and such stones as the "rain-bringing stone," the "shaving stone," the "cold-bringing stone," horn, glass, etc. With al Tifaschi's book *Azhar al Afkar fi Djawahir al Ahdjar* (Best Thoughts on the Best of Stones), written in the middle of the thirteenth century, the science of gemology was already firmly rooted in the Arab world. This book revolves around al Tifaschi's work and is based on the edition of Dr. M. Y. Hassan and Dr. M. B. Khafaga, who compared, verified, amalgamated, and unified six of the many existing versions of al Tifaschi's manuscript, including one original kept in *Dar al Kutub al Masriyya* in Cairo, Egypt. As for Ibn al Akfani, he gathered all the "elders" work in one short concise book called *Nukhab al Zakha'ir fi Ahwal al Djawahir* (meaning roughly Selection of Best Knowledge on Precious Stones), written in the middle of the fourteenth century.

In addition, the following books were written on precious stones:

• *Al Durar al Kamina* (The Hidden Precious Stones), author and date unknown.

• *Kinz al Tudjar fi Ma'arifat al Ahdjar* (The Treasure of Merchants with regards to the Knowledge of Gemstones). This manuscript is in the Bibliothèque Nationale in France and was written in 1282 by Bailak al Qabadjaqi in Cairo. Twenty-three Greek and Arab authors are mentioned therein, including Theophrastus, Aristotle, Pliny, Hermes, Ptolemy, al Mas'udi, al Ghazali, al Biruni, and al Tifaschi. This book devotes lengthy passages to the talismanic and medicinal properties of gemstones.

• *Rasa'il ba'ad al Hukama wal 'Ulama al Qudama' fil Djawahir wal Khawass* (Epistles Written by some Ancient Philosophers and Scholars on Gemstones and their Properties). This book is attributed to Ibn Sina (Avicenna).

All the above authors wrote books devoted exclusively to precious stones, except for al Mas'udi, who wrote a sort of large history, travel, geography and other subjects encyclopedia, incorporating in it articles on gemstones, their occurrence, localities, and benefits.

The most important and prolific writer on gemology seems to have been Djabir al Kharaqi (or Khiraqi?), (exact date unknown, but must be eighth century as he was mentioned in Ibn al Nadim's *Fihrist*, the bio-bibliography written in the eleventh century, as being a student of Djabir ibn Hayyan). Al Kharaqi wrote the following long series of books devoted exclusively to precious stones:

- *Kitab al Ahdjar al Awwal* (Book on Gemstones, volume 1)
- *Kitab al Ahdjar al Thani* (Book on Gemstones, volume 2)
- *Kitab al Djawahir al Kabir* (The Great Book of Gemstones, which could also have been called The Great Book on Pearls or The Great Book on Jewelry as the word *Djawahir* in Arabic refers to all the above three terms).
- *Rasa'il fil Hadjar* (Epistles on Gemstones)
- *Kitab al Zumurruda* (A Book on Emerald)
- *Kitab al 'Aqiqa* (A Book on Cornelian)
- *Kitab al Billawra* (A Book on Rock Crystal)
- *Kitab al Sate'i* (A Book on Diamond).[2] *Al Sate'i* in Arabic means "the brilliant," and as the author had already written a book on rock crystal, this brilliant stone could only have been diamond.

It is most unfortunate indeed that none of these books has survived, unless they are still in manuscript form somewhere. According to the King Faisal Center for Research and Islamic Studies in Saudi Arabia, it is estimated that in Turkey alone there are approximately one million manuscripts, only a quarter of which are indexed and cataloged. At any rate, it is estimated that over fifty lapidaries were composed by medieval Arab authors.[3]

Well-Known Gemstones in the Thirteenth Century

English name	Old Arabic name	Phonetic equivalent in English
Pearl	الجوهر	*Djawhar*
Ruby	الياقوت	*Yaqut*
Sapphire	الياقوت الاسمانجوني	*Yaqut asmandjuni*
Emerald	الزمرد	*Zumurrud*
Peridot	الزبرجد أو الزبردج	*Zabardjad* or *Zabardadj*
Spinel	البلخش	*Balakhsh*
Garnet	البنفش	*Banafsh*
Almandine garnet	الماذنج	*Mazendj*
Tourmaline	البجادي أو البزادي	*Badjadi, Bijadi* or *Bazadi*
Diamond	الالماس أو مــاس	*Almas* or *Mas*
Moonstone	حجر القمر	*Hadjar al Qamar*
Moonstone cat's eye	عين الهر	*'Ain al Hirr*
Turquoise	الفيروزج	*Fairuzadj*
Cornelian	العقيق	*'Aqiq*
Agate	الجزع	*Djaz'*
Magnetite	المغناطيس	*Maghnatis*
Emery	السنباذج	*Sunbazadj*
Malachite	الدهنج	*Dahnadj*
Lapis lazuli	اللازورد	*Lazuward* or *Lazward*
Coral	المرجان	*Mardjan*
Jet	السبج	*Sabadj*
Amethyst	الجمشت	*Djimashtt*
Jade	اليشم	*Yashm*
Jasper	اليصب	*Yassb*
Rock crystal	البلور	*Billawr* or *Ballur*
Talc	الطلق	*Talq*
Amber	الكهربا	*Kahraba*
Zircon	الجرقون أو الخرجون	*Djargun* or *Kharjun*
Bezoar	البازهر	*Bazahr*

Modern Arabic name	Phonetic equivalent in English
اللؤلؤ	*Loulou* or *Lou'lou'*
الياقوت	*Yaqut*
السفير	*Safeer*
الزمرد	*Zumurrud*
الزبرجد	*Zabardjad*
السبينل أو البلخش	*Spinel* or *Balakhsh*
الجارنت أو البنفش	*Garnet* or *Banafsh*
الجارنت أو المازنج	*Garnet* or *Mazendj*
البجادي	*Bijadi*
الالماس أو ماس	*Almas* or *Mas*
حجر القمر	*Hadjar al Qamar*
عين الهر	*'Ain al Hirr*
الفيروز	*Fairuz*
العقيق	*'Aqiq*
الجزع أو الاغات	*Djaz'* or *Agate*
المغناطيس	*Maghnatis*
السنباذج	*Sunbazadj*
الدهنج	*Dahnadj*
اللاذورد	*Lazuward,*
اللاذورد أو اللابيس	*Lazward* or *Lapis*
المرجان	*Mardjan*
السبج	*Sabadj*
الجمشت أو الامتيست	*Djimashtt* or *Ametist*
اليشم أو الجاد	*Yashm* or *Djad*
اليصب	*Yassb*
البلور	*Billawr* or *Ballur*
الطلق	*Talq*
الكهرمان أو الكهربا أو الكوربا	*Kahraman, Kahraba* or *Koraba*
الزرقون	*Zircon*
البازهر أو البنزهير	*Bazahr* or *Banzahir*

The World Where al Tifaschi Lived

THE beauty and mystery of gemstones have captured people's interest since antiquity. It did not take long for legends, occult powers, and medicinal purposes to be attributed to these stones, and no powerful king or ruler would let his coffers go empty of these rare and valuable gems.

The Greeks were the first to write about precious stones and were often quoted by the Arabs who followed them closely in this field. All the holy books mention gemstones, proof of their special value and consideration amongst all nations. For example, ruby, rock crystal, and pearl are mentioned in the Qur'an. The word *Mardjan* is also mentioned, but it then referred to small (seed) pearls and not to coral, as it is the case today. However, the science of gemology based on scientific tests, geological finds, and acute observation with the aid of special instruments is a relatively young one: Only in the sixteenth century were the first scientific books on gemstones written.

At least, this is what we know so far.

The study of some manuscripts dating from the ninth to fourteenth centuries written by Arab scholars tells us otherwise, for, at least by the year A.D. 800 (and probably before; unfortunately no such writings have survived), books were written on precious stones, that can truly be considered the first scientific basis of gemology.

Several Arab authors have written treatises on gemstones, some of which did not survive; others are lying still undiscovered or unresearched in neglected archives and libraries. Amongst the most famous Arabs who wrote on precious stones and whose books have survived are: Al Kindi (ninth century), who inherited his knowledge on gemstones from his grandfather Sabah al Kindi; al Biruni (tenth-eleventh century), al Tifaschi (thirteenth century) and Ibn al Akfani (fourteenth century).

At the beginning of the eighth century, and less than a century after the death of the Prophet Mohammad in A.D. 632, the banner of Islam extended west from Spain and North Africa to as far east as Persia, India, and the borders of China and south to the tip of the Arabian peninsula. By the thirteenth century, Arab expansion covered Spain, North Africa, Egypt, Syria, Turkey, Persia, Arabia, and India. During this time, Europe was busy preparing for the Crusades, which lasted from 1035 to 1395. The American continent was not discovered until 1492, and, until its discovery in 1788, Australia was still unknown. The first attempt by the West to discover and record the mysteries and cultures of the East was the inland voyage of Marco Polo, who went as far as China between the years 1245 and 1323.

The ninth to fourteenth centuries witnessed the golden age of Arab civilization. In the ninth century, two names stood "supreme in world affairs: Charlemagne in the West and Harun al Rashid in the East. Of the two, Harun was undoubtedly the more powerful and represented the higher culture." [4]

Encouraged by their enlightened rulers, scholars were prolific in various fields, such as literature, poetry, biology, medicine, astronomy, and mathematics. It was also a period of great military powers and successes, which witnessed the rise to power of the Mamluk dynasties in Egypt, the Umayyads in Cordoba, Spain, the 'Abbasid Caliphate with capitals in Baghdad and Samarra' in Iraq, and the Seljuks in Persia, Mesopotamia and Turkey. (The latter were known as Seljuks of Rum.)

Commerce was also flourishing at that time, and the exchange of goods between the West and the East was regular and abundant. People often migrated from al Andalus (Spain) to live in the peaceful and stable countries in the Middle East and North Africa. It was common practice for scholars to write on different interrelated subjects in the same book. This does not come as a surprise to us, as today, also, the science of gemology for example, overlaps several other sciences, including mineralogy, biology, zoology, and medicine, exactly as was the case in the ninth to fourteenth centuries.

Ahmad ibn Yusuf al Tifaschi

HIS full name is Shihab al-Din Abul 'Abbas Ahmad ibn Yusuf al Tifaschi. He was born in Tifasch (at that time, part of Tunisia, now part of Algeria), in A.H. 580 (A.D. 1184). He studied first in Tifasch, then in Tunis and visited Cairo and Damascus where he stayed long periods of time, studying under well-known scholars of these countries. He worked as a judge both in Tunis and in Cairo, where he spent most of his life. He traveled several times to Iraq, Armenia, and Persia (then called *Bilad al Fors*), seeking knowledge on minerals and gemstones.

Al Tifaschi came from an important, wealthy family whose members had long occupied key posts, as judges, ministers and governors. His own father was a judge by profession, but he was also well acquainted with gemstones, and he communicated his knowledge of gemstones to his son. Like his uncle, Ahmad al Tifaschi was also a scholar and a poet: He wrote many books on various subjects, including his famous book on gemstones: *Azhar al Afkar fi Djawahir al Ahdjar* (Best Thoughts on the Best of Stones). He also wrote an early form of encyclopedia, where he gathered all his knowledge and experience. According to Abdel Qadir Zamamah, the only surviving copy of this encyclopedia was purchased by Ibn Manzur, a contemporary of al Tifaschi, and an extract of poems was selected and rewritten in more precise and concise form with another title, under Ibn Manzur's own name. Some of al Tifaschi's other books dealt with the following subjects: Medicinal values of stones, sexual knowledge, poetry of the thirteenth century, medicine, and an atlas of the countries known at that time.

Al Tifaschi was described by some western scholars as being a gem dealer, by others as a jeweler. However, his main occupations were those of judge, poet, and scholar. His knowledge of gemstones was inherited from his father and perfected through broad personal experience, intensive reading of many important ancient and medieval books written on the subject, and extensive travel to trading and mining centers of gemstones at that time. Al Tifaschi wrote his book on gemstones approximately ten years before his death in Cairo in A.H. 631(A.D. 1253), where he was buried. According to Clément Mullet, *Azhar al Afkar fi Djawahir al Ahdjar* was the most "methodical and complete"[5] work on precious stones.

Al Tifaschi's broad and encompassing education is obvious from his writings on so many different subjects. Moreover, he was familiar with the writings of old Greek, Latin, Persian, and Arab scholars, quoting Aristotle, Theophrastus, Pliny, Dioscouridis, and others many times in his book. The fact that al Tifaschi was familiar with Theophrastus's work on stones leads us to assume that:

1. There were Arabic translations of the works of Greek scholars readily available by at least the ninth century. For example, J. Ruska translated into German the Arabic version of Aristotle's manuscript on stones, *Kitab al Ahdjar li Aristatalis,*[6] which he dated to sometime before the middle of the ninth century. However, J. D. Rouse[7] stated that Aristotle did not write on stones. Therefore, the book on gemstones attributed to him was written either by someone else called Aristotle or by someone using the name.

Whether this person was Arab or one familiar with the language is still unknown. As a matter of fact, Abul Raihan al Biruni was the first to make a remark in his book entitled *Al Djamahir fi Ma'arifat al Djawahir* (General Knowledge on Gemstones)[8] that the book on gemstones was wrongly attributed to Aristotle. This remark by a tenth-eleventh-century Arab scholar considered to be one of the most brilliant and knowledgeable of Arab scholars supports Mr. Rouse's assumption. Several sentences in the book stating that: "My pupil Alexander . . ." or "My pupil Alexander was the first one to discover the mining sites of this stone . . ." could lead the reader to assume that the book was indeed written by Aristotle. However, in other instances, the explanation of the origin of the word "bezoar" as originally coming from the Greek "antipoisonous" while in fact it is from the Persian, and the prices often quoted in *Dirhems* would refute this argument. At any rate, if the above book was indeed written by an Arab author, then other copies of the Paris manuscript called *Kitab al Ahdjar li Aristatalis* mentioned above could still have survived somewhere in the Arab or Islamic world. As a matter of fact, the Arab scholar Father Anastase Marie de Ste. Elie[9] stated that Aristotle's book on gemstones was translated by a certain Luqa Ibn Serafion (or Asrafion?). Therefore this book could also have been written by another Greek scholar using the pseudonym Aristotle or by Luqa Ibn Serafion himself.

2. Another possibility is the following: As al Tifaschi quoted Pliny on several occasions in his book, giving us the title of Pliny's book in Arabic, *Sirr at-Tabi'at fil'Ilal wal Ma'alulat* (Nature's Secrets on Causes and Effects) by which he must have meant Pliny's *Natural History*, he could have obtained his information on Theophrastus from Pliny (who referred to Theophrastus), and then quoted it directly in his book omitting reference to Pliny. Al Tifaschi quoted Theophrastus as saying that rock crystal can melt like glass and can easily be dyed.[10] Theophrastus did mention in passing the fusibility of quartz; however, Pliny did not elaborate on this point. Therefore, it would sound more probable that al Tifaschi was acquainted with Theophrastus's work firsthand.

3. Al Tifaschi could have been well versed in Greek and Latin, which enabled him to study and quote the above mentioned Greek and Latin scholars. He often tells that So–and–so "mentioned in his *Book on Stones*," or "the old sages advised . . ." etc. This could very well be the case, in view of al Tifaschi's extensive education and knowledge in so many different subjects, and also taking into consideration his important and intellectual family background, as well as his highly respected social

position. Let us also not forget that he lived in several important Arab cities at that time, Tunis, Cairo, and Damascus, and that his travels led him as far as Armenia and Samarqand. If indeed this is the case, then there could have been a manuscript in Greek on the *Treatise on Stones* by Theophrastus available in the Islamic world. This could, for example, confirm that Codex Vaticanus Graecus 1302, one of the three surviving manuscripts of Theophrastus's *Treatise on Stones* dates indeed from the twelfth century as suggested by Messrs. Devreese and Gianelli.[11]

Al Tifaschi had several "European" friends; he himself mentioned on different occasions: "One of the European merchants gave the king of Africa . . ." and "I was visiting the shop of a gem expert from al Andalus . . ." (Spain); and he often referred to *Bilad al Afrandjah* (Europe) and *Bilad al Rum* (Territory of the Romans, Asia Minor). He must also have been a frequent visitor to the Venetian settlement that existed in Aleppo, Syria, until the middle of the thirteenth century. Hence, a similar copy of this manuscript, dealing with gemstones, could have attracted his attention, and he could have obtained a copy.

At any rate, the fact that al Tifaschi mentioned the Arabic title of Pliny's book proves that Arabic translations of major Greek and Latin works were readily available by at least the ninth century (before al Kindi's time, who was also familiar with them).

As a matter of fact, a great movement of translation had started in the Arab world soon after the death of the Prophet Mohammad and was continued during the Umayyad and 'Abbasid periods. During the Umayyad period A.H. 40-132 (A.D. 660-750), Khalid ibn Yazid ibn Mu'awiyah, considered to be one of the founders of the translation movement in the Islamic world, had developed his grandfather's well-known library called *Beit al Hikma*, or *Dar al Hikma* (House of Wisdom) into a large center for books and translations on all fields of science. The 'Abbasid period that followed A.H. 132-656 (A.D. 750-1258) was known to be the era of translations in the Islamic world. The *Khalifas* al Mansur and Harun al Rashid (eighth-ninth century) sponsored and developed this movement, and one of the greatest scholars and most prolific translators of the age was Hunayn ibn Ishaq (Joannitius, 809-73), who was appointed by the *Khalifa* al Ma'mun superintendent of his library-academy. His son Ishaq became the greatest translator of Aristotle's works.[12]

Different from Pliny, al Tifaschi's account is based on pragmatic experience, firsthand observation, and scientific testing of gemstones, gathered from his own personal experience and familiarity with gemstones. Indeed, al Tifaschi was the first one to use modern, "scientific" terms when describing certain characteristics of gemstones, including such words as: Alluvial deposits (*hasba'a*), artificial stones (*ahdjar mudallasa, masnu'a*), cleavage (*inshiqaq, tashqiq*), false cleavage (*shibh tashqiq*), flame test (*ikhtibar al djamr all'azi laysa lahu dukhan*), streak powder (*hukaka*), internal stress or cracks (*tash'eer, sha'rat*), pleochroism (*ikhtilaf al sibgh*).

When describing moonstone cat's eye, al Tifaschi said: "I did not find any mention of this stone in any book on stones whether old or new, I think it has come recently in people's hands . . ."

From the above, we can assume that al Tifaschi conducted extensive research on the books written about gemstones available at that time before writing his own book. He gathered information from the most important earlier Arab and classical authors who wrote about precious stones, adding his own valuable experience and knowledge. His major contribution, however, is to have described the first scientific method to study gemstones, treating all their physical qualities (hardness, specific gravity (weight), dispersion, luster, brilliance, quality, color, crystallography, habits, inclusions, etc.) in addition to listing their average prices by weight/per carat, ways to differentiate between genuine and imitation stones, some chemical tests used to manufacture stones and test them; their medicinal benefits, occurrence, and localities, all within a scientific framework divided into clear categories that he himself described as follows: "The stones described [in this book] are twenty-five in number, namely: pearl, corundum, emerald, peridot, spinel, garnet, tourmaline, diamond, moonstone cat's eye, bezoar, turquoise, cornelian, agate, magnetite, emery, malachite, lapis lazuli, coral, jet, amethyst, hematite, jade, jasper, rock crystal, talc. We shall talk about each one of the above-mentioned stones after explaining the origin of its Arabic name, according to the following five aspects: First, cause of formation of the stone in its host rock; second, characteristics of the mine where it was formed; third, good and bad qualities, also genuine and fake stones; fourth, properties and benefits of the stone; fifth, general average value and prices."[13] While ancient and medieval systems of classification of minerals were based primarily on the appearance of stones, al Tifaschi's system was different, based on scientific principles and value of gemstones.

Regarding "benefits" of stones, al Tifaschi devoted a paragraph in each chapter to the healing powers and talismanic properties attributed to the stone. Here we find the theme of magic and medicinal uses that started from the ancient period, even before Theophrastus and Pliny, and lasted well into the eighteenth century. The Arabs, however, molded gemstone therapies into a proper medical sytem; and we note that al Tifaschi's reports were actually limited to medical uses and a few superstitions of the time. Taking into consideration that, during these times, the "benefits" that a stone could provide substantially increased its value, and that medieval medicine was based first and foremost on herbal, gemstone, and mineral therapies, we can understand that al Tifaschi had no choice but to incorporate a section on gemstone therapies in his otherwise scientific book. Clément Mullet thought that by the word *khawass* (literally meaning properties), al Tifaschi meant the "talismanic properties" of the stone. However, al Tifaschi used the word *khawass* as a title for paragraphs dealing with characteristics of the stone (both external and internal), medicinal value, and usage of the stone in physical and spiritual healing, as well as talismanic properties.

Al Tifaschi has indeed presented us with a scientific treatise on gemology that set up the methodology for later lapidaries, well before the "father of mineralogy" Georgius Agricola, whose work *De Natura Fossilum* appeared in 1546; or Anselm Boetius de Boodt, whose work *Gemmarum et Lapidum Historia* (1609) is considered to be a landmark in lapidaries.

Notes

1. Hitti, *History of the Arabs*, 240.
2. Ibn al Nadim, *Fihrist,* 355.
3. Hitti, 383.
4. Ibid., 298.
5. Mullet, "Essai sur la Minéralogie Arabe," *Journal Asiatique*, 5.
6. Encyclopedia of Islam, 250.
7. Rouse, *Garnet,* 16.
8. Al Biruni, *Al Djamahir fi Ma'arifat al Djawahir*, 41.
9. Ibn al Akfani, *Nukhab al Zakha'ir fi Ahwal al Djawahir*, 21.
10. Al Tifaschi, *Azhar al Afkar fi Djawahir al Ahdjar,* 81.
11. Caley and Richards: *Theophrastus on Stones*, commentary, 5.
12. Hitti, *History of Arabs*, 312 - 313.
13. Al Tifaschi, 2.

CHAPTER 2

Gemological Terminology Used by Ahmad al Tifaschi in the Thirteenth Century

English word	Corresponding Arabic word used by al Tifaschi	Phonetic equivalent in English
Air bubbles	ريح	*Rih*
Alluvial deposits	الحصباء	*Hasba'*
Artificial stones (or cheated)	أحجار مدلسة	*Ahdjar mudallasa*
Man-made stones	أحجار مصنوعة	*Ahdjar masnu'a*
Brilliance, refractive index	الشعاع	*Shu'a'a*
Brittleness	الرخاوة	*Rakhawa*
Calcinate	يتكلس	*Yatakallas*
Carat	قيراط أو قرّاط	*Qirat* or *kirrat*
Cat's eye	عين الهرّ	*'Ain al Hirr*
Chatoyancy	كأنها عين الهرّ المتحرك في مقلته	*Ka'annaha 'Ain al Hirr al mutaharrik fi muqlatihi*
Cleavage	إنشقاق، تشقيق	*Inshiqaq, tashqiq*
Cleavage, false	شبه تشقيق	*Shibh tashqiq*
Color	اللون، الصبغ	*Lawn, sibgh*
Crystal	القصب، القصبة	*Qasab, qasaba*
Decrepitates	ينحل على النار	*Yanhall 'ala'l nar*
Defect	عيب	*'Ayb*
Flame test	إختبار الجمر الذي ليس له دخان	*Ikhtibar al djamr al 'lazi la lahu dukhan*
Grinding	المحك	*Mahakk*
Hardness	الصلابة	*Salaba*
Heat treatment	معالجة بالنار	*Mu'alaja bil'nar*
Imitation stones	أشباه	*Ashbah*
Inclusions (in terms of defects)	عيوب	*'Uyub*
Internal stress or cracks	التشعير، الشعرة	*Tash 'eer, shu' rat*
Liquid (inclusion)	ماء	*Ma'*
Luster, clarity	المائية، الماء	*Ma'iyyah, ma'*
Matrix	تراب المعدن	*Turab al ma'dan*
Mean value	القيمة الوسطى	*Qima al wusta*

Mother rock	المعدن	*Ma'dan*
Placer deposits	الحصباء	*Hasba'*
Pleochroism	إختلاف الصبغ	*Ikhtilaf al sibgh*
Polishing	الجلاء	*Djala'a*
Rutile needles	السوس	*Souss*
Scratch test	الخدش، المخدش	*Khadsh, makhdash*
Sheen	إشراق، بصيص	*Ishraq, Basiss*
Small stone (suitable for a ring)	الفص	*Fass*
Sphericity	الدحرجة	*Dahradja*
Streak, streak power	الحكاكة	*Hukaka*
Specific gravity	الوزن	*Wazn*
Tranparency	الشفوف، شفافة	*Shufuf, shafafa*
Supply and demand	طلب السوق والعرض	*Talab al suq wal'ardh*
Vein	العِرق	*'Irq*
Working the stone	الهندمة	*Handama*

THE aforementioned table is important in that it shows that distinct gemological terms were used as early as in the thirteenth century when describing characteristics of gemstones.

Nowadays, it is common practice to classify information on gemstones according to the following guidelines: Name, chemical composition, color, color of streak, hardness, specific gravity, cleavage, transparency, luster, dispersion, occurrence (host mineral), and locality. It is amazing to note that al Tifaschi used exactly the same classification in his book on gemstones. In fact, he was probably the first person to use in his book devoted exclusively to gemstones the exact modern classification mentioned above.

Al Tifaschi characterized his book as being unique in its kind and of great benefit, describing the precious stones that "no great king or important leader can do without in view of their unusual benefits and great properties."[1]

The order in which al Tifaschi described the stones is as follows: pearl, corundum, emerald, peridot, spinel, garnet, tourmaline (rubellite), diamond, moonstone cat's eye, bezoar, turquoise, cornelian, agate, magnetite, emery, malachite, lapis lazuli, coral, jet, amethyst, hematite, jade (nephrite), jasper, rock crystal, and talc. Diamond was the most valuable stone for the Romans and in earlier times in India, while the Persians classified diamond in fifth place after pearl, ruby, emerald and chrysolite.[2]

The Arabs held pearl and ruby paramount, and one could assume that al

Tifaschi's classification was not due solely to chance, as color was the major criterion upon which the value of stones was based, and diamond was well behind pearl, ruby, emerald, and spinel in the order of importance at that time.

After explaining the origin of the Arabic name, al Tifaschi wrote about each one of the stones according to the following:

1. Cause of formation of the stone in its host rock
2. Characteristics of the mine (location) where the stone was formed
3. Good and bad qualities; genuine, imitation, and fake stones
4. Characteristics and benefits of the stone
5. General average values, and prices.[3]

Under the characteristics of the stone, al Tifaschi mentioned the crystallographic forms of some stones (e.g., diamond); the chemical composition of others (e.g., amethyst, malachite, turquoise, hematite, and magnetite); color, hardness, weight (specific gravity), cleavage (in diamond), streak, streak powder, transparency, luster, brilliance, dispersion, and species to which some stones belong, such as coral, which, according to him is midway between the plant and mineral kingdoms. Although al Tifaschi did not mention every one of the above characteristics under each stone, he did mention them when it was important, such as in the case of crystal habit and cleavage of diamond. This is probably the earliest record of such a classification of stones, upon which our modern books are based.

Through observation, empirical evaluation and simple tests, early Arab scientists were able to make discoveries and advance theories whose accuracy sometimes amazes the modern gemologist. For example, how could Ahmad al Tifaschi ascertain in the thirteenth century that copper entered in the chemical composition of turquoise? It is only recently (in the twentieth century) that copper was determined to be responsible for the beautiful, perfect sky-blue, color of turquoise, although R. Webster still suggests that "the cause of the beautiful blue colour of turquoise is not known for certainty."[4] How did al Tifaschi guess that amethyst owed its color to iron? This was only discovered later in the twentieth century, as R. Webster again points out: "Thanks to recent researches and to the successful synthesis of amethyst by hydrothermal methods, it is now certain that the amethyst hue is due to radiation induced colour centres associated with an iron impurity."[5] Furthermore, while spinel was recognized in the West to be a different species from ruby a mere 150 years ago,[6] al Tifaschi already knew this stone had nothing to do with ruby, except perhaps their similarity of color, which he says can lead the person to confuse it with the precious ruby. As a matter of fact, the Arabs knew that spinel was a different species from ruby as early as the year A.D. 800, that is more than 900 years before the West realized this fact. This is confirmed in the writings of al Biruni, who states in his *Al Djamahir fi Ma'arifat al Djawahir*[7] that spinel was confused with ruby until the times of Al Bouayh (the Buyids, whose rule started in the tenth century), when they realized it was a different species and priced it at half the price of ruby. Also, while in the West sapphire and ruby were discovered to belong to the same group (the corundum group) only in the nineteenth century,[8] this was already an established fact in the thirteenth century, as stated by al Tifaschi in his book. Al Tifaschi correctly described and classified the various types and shades of corundum, mentioning specifically that ruby was the most valuable of the group, followed by sapphire. The orange-pink variety of corundum (known nowadays as padparadshah) was considered the most valuable of the yellow to orange group, exactly as is the case today. Each of the above points will be discussed at length in the individual chapters on stones.

Hardness of Gemstones

APART from the outstanding achievements already mentioned, perhaps the most important contribution al Tifaschi made is one that is considered the first pillar of the science of gemology as it began in the nineteenth century. In the third century B.C., Theophrastus wrote that some stones "can be carved with iron, but only with rather blunt tools."[9] He was probably one of the first to point out different hardnesses in stones. Pliny and other ancient authors continued to note this characteristic, stating that "there is such a vast diversity in their degree of hardness, that some stones do not admit of being engraved with iron, and others can only be cut with a graver blunted at the edge. In all cases, however, precious stones may be cut and polished by the aid of adamas."[10]

However, the landmark in this question was the Mohs scale of hardness: In 1822, a German scientist, Friedrich Mohs, devised the first scale by which stones could be tested for hardness: the Mohs scale of hardness. This scale is based on the fact that a harder material will scratch a softer one. He selected ten minerals, each of which will scratch the mineral lower in hardness but not the one higher on the scale:

Talc	1 (softest stone on the scale)
Gypsum	2
Calcite	3
Fluorspar	4
Apatite	5
Feldspar	6
Quartz	7
Topaz	8
Corundum	9
Diamond	10 (hardest of all substances)

This scale is nonquantitative in that the difference in hardness between diamond at number 10 and corundum at number 9 is greater than that between talc at number 1 and corundum at number 9.

However, more than 600 years earlier, al Tifaschi had devised the following scale of hardness: He stated that diamond was the hardest of all stones and gave it the arbitrary number of 4. He then described corundum as the stone next in hardness after diamond, giving it the number 3 on his scale together with emery, followed by emerald, spinel, garnet and tourmaline, which, according to him, were all hard stones, although softer than corundum. He described turquoise as a medium-soft stone, although harder than malachite and lapis lazuli, and said that agate was one of the most resistant stones, giving it the number 2. Coral was a soft stone, with number 2; talc was also placed at number 2, although he described it as being a "hard" stone (meaning flexible and highly resistant to heat and fire); and jet was one of the softest stones, at number 1 on his scale.

Al Tifaschi's scale is thus established as follows:

Diamond	4 (hardest of all stones)
Corundum, emery	3
Garnet, emerald, spinel, tourmaline	2 3/4
Turquoise	2 1/2
Malachite, lapis lazuli	2 1/4

Coral, talc, agate	2
Jet	1

Agate was a hard stone, difficult to work, according to al Tifaschi, although he gave it the hardness of only 2 on his scale.

Below are the modern hardness numbers of the stones described by al Tifaschi:

Talc	1
Jet	2 1/2 - 3 1/2
Pearl	2 1/2 - 3 1/2
Coral	3 1/2 - 4
Malachite	4
Lapis lazuli	5 1/2
Turquoise	5 1/2 -6
Hematite	5 1/2 - 6 1/2
Moonstone	6
Garnets	6 1/2 - 7 1/2
Jade (incl. jadeite and nephrite)	6 1/2 - 7
Quartz (incl. rock crystal, amethyst, agate, jasper)	7
Tourmaline	7 - 7 1/2
Emerald	7 1/2 - 7 3/4
Spinel	8
Corundum, emery	9
Diamond	10

The following table lists the hardnesses attributed to stones by al Tifaschi versus Mohs:

Gemstone	al Tifāschi	Mohs
Jet	1	2 1/2 - 3 1/2
Talc	2	1
Coral	2*	3 1/2 - 4
Malachite	2 1/4	4
Lapis lazuli	2 1/4	5 1/2
Turquoise	2 1/2	5 1/2 - 6

* Hardness number 2 is taken from the Cairo version. In this version, it is written that coral is hard in the third degree.

Agate	2	7
Tourmaline, garnet, emerald, Spinel	2 3/4	7-7 1/2; 6 1/2- 7 1/2; 71/2 -7 3/4 and 8 respectively
Corundum, emery	3	9
Diamond	4	10

Although the hardness number is not the same in both tables, and except for talc and agate, the order of the softest to the hardest stone is correct and although al Tifaschi did not elaborate on the difference in hardness between tourmaline, garnet, emerald, and spinel, his results were remarkably accurate and in accord with our modern estimates of the comparative hardness of stones. Al Tifaschi's hardness scale is indeed the basis of the Mohs scale of hardness.

As early as the ninth century, al Kindi used two methods to test gemstones: the heat treatment test and the scratch test. It is also worth mentioning that another Arab gemologist of the fourteenth century, Ibn al Akfani, wrote in the emerald chapter that "emerald is tested with a pointed fragment of *'Aqiq* (cornelian). If it scratches it, then the stone is only one that resembles emerald."[11] Thus, the idea of a scratch test to determine the hardness of stones and differentiate between them was used by the Arabs since the ninth century, and cornelian, with a hardness of 7 (being a variety of quartz), would not scratch a genuine emerald (which has a hardness of 71/4-7 3/4). Even the relatively slight difference in the hardness of both stones was enough to differentiate between them with the scratch test at that time.

Al Tifaschi established another scale, which he called the scale of heat and cold:

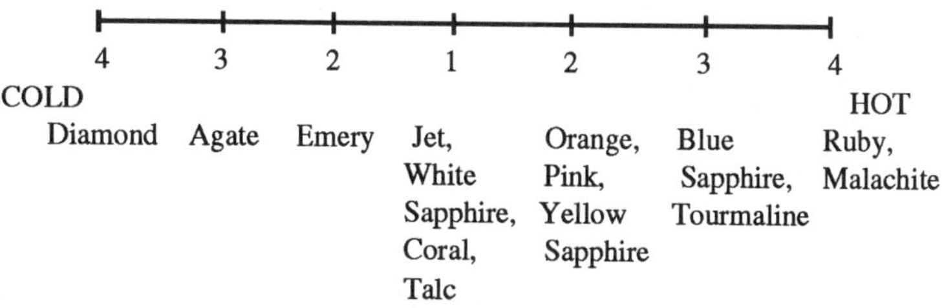

This theme of heat and cold was taken from the ancient authors (such as Theophrastus and Pliny) and was possibly related to color (or its lack). Clément Mullet explains it thus, basing his explanation on al Tifaschi's theories: "If, (during the initial stage of the formation of the a stone) . . . heat and drought prevail, the stone will be red: It is a ruby. If heat weakens, the stone is yellow: It is a topaz. If heat is temperate and moderate, the stone is white: It is a colorless ruby. If drought and cold prevail, a black shade results. Sometimes this (black) shade is only superficial and the interior (of the stone) remains red. Sometimes also the red and black shades combine at the surface and produce the blue color."[12]

Gemstones Not Mentioned by al Tifaschi

THE reader will note that several stones considered valuable nowadays were not mentioned by al Tifaschi. These include: opal, aquamarine, morganite, heliodor, alexandrite, chrysoberyl, some members of the garnet family, topaz, zircon, and amber.

Two reasons explain this omission:

1. Only those stones considered most precious at that time were mentioned, the common ones of no medicinal or other value and the rare, unfamiliar stones being omitted,[13] as al Tifaschi himself clearly states in the introduction of his book.
2. As noted, the banner of Islam at that time extended from Spain and North Africa to Turkey, westward to India, and south to Arabia. Western Europe was known as *Bilad al Afrandj* or *Afrandjah*; the Byzantine Empire as *Bilad al Rum* (Territory of the Rum in Asia Minor). The North American continent was discovered only in 1492 and Australia in 1788. Southwest and South Africa were unfamiliar. Therefore those stones that are found mainly in America, Australia, Central, Southwest and South Africa were not known in the ninth to fourteenth centuries; they included opal, aquamarine, morganite, heliodor, alexandrite, cat's eye (cymophane), and some members of the garnet family.

Nowadays, the Arabic name for cat's eye (*'Ain al Hirr*) usually refers (erroneously) to tiger's eye or hawk's eye (respectively, the golden-brown and bluish-brown crystalline varieties of quartz displaying chatoyancy).

'Ain al Nimr (tiger's eye) is also used to describe the tiger's eye variety. Several members of the garnet family were already known by the thirteenth century, namely, almandine, pyrope-almandine, spessartine, and hessonite. Other garnets are either not gem varieties or have been only recently discovered in rare and small specimens. Topaz, zircon and amber were also not mentioned by al Tifaschi. This could be because they were mistaken for another stone similar in appearance and variety of colors, or they were not considered of any importance (medicinal or otherwise), as it could be the case with amber.

Regarding amber, many people in the West think that the Arabic name of amber is *'Ambar*. This is not correct, as *'Ambar* refers to ambergris, which is a secretion from the intestines of whales and is used for making expensive perfumes. Amber is called *Kahraman* or *Kahraba* in Arabic *(Koraba* in colloquial Arabic), which means electricity. Amber was known at least from the times of al Kindi (ninth century), and al Biruni mentioned that this stone was called *Kahraba* because it attracted straw, feathers, and the like when rubbed against a piece of cloth or a beard, and that the *Rum* (Byzantines) people called it "electron."[14]

As for zircon, it could have been mistaken for a corundum variety, especially because al Tifaschi mentioned that there are two types of white corundum: One of strong white color, high luster and brilliance, and another of lesser brightness, brilliance, and luster, and heavier in weight than the other corundums. If we compare the specific gravity of zircon (4.4-4.69) with that of corundum (3.96-4.01), we note that zircon is indeed heavier in weight and therefore could have been thus misidentified. As a matter of fact, a stone called *Kharjun* or *Khargun* was mentioned by al Biruni as being similar to rubellite. This could also have been *Zargun* (zircon), as al Biruni's handwriting was bad, and his manuscripts were hardly readable.

Regarding topaz, it is worth mentioning that in the West, peridot was called topaz in the old times (perhaps in reference to the island of Topazios in the Red Sea, also called *Zabardjad* island or St. John's island). Some confusion exists about this stone, as peridot was called *Zabardjad* in Arabic (at least from the year A.D. 1240 onwards), and this word always referred to green-colored stones. Before the year 1200, *Zabardjad* used to refer to both emerald and peridot, not surprisingly, as the only source for both stones in older times was Egypt.

We can therefore conclude that, apart from those stones found exclusively in Australia, America, and Central and South Africa, the well-known gemstones of the ninth to fourteenth centuries were almost exactly the ones today considered the most valuable ones, namely, pearl, diamond, corundum (including ruby and sapphire), emerald, spinel, garnet, turquoise, lapis lazuli, jade, peridot, tourmaline, moonstone, quartz (including rock crystal, amethyst, jasper, and agate), coral, malachite, jet, and hematite, as well as magnetite, talc and emery; the latter ones for their uncommon and valuable uses.

In the West, the twentieth-century concept of a "precious stone" is based on Jerome Cardan's definition which he established in the sixteenth century.[15] Cardan classified all stones that appeared brilliant as gems. The term "precious stones" was reserved for those that were rare and small as well as being brilliant. The Arabs called gemstones *al Ahdjar al Karima*, meaning stones that are noble, valuable, expensive, rare, and precious at the same time. The rarest and most valuable were also called *al Ahdjar al Mulukiyya*, or royal stones. According to the *Encyclopedia of Islam*,[16] the word *Djawhar* (plural *Djawahir*) had the meaning of "jewel" (and "jewelry") from at least the middle of the ninth century, as was mentioned in the Paris manuscript *Kitab al Ahdjar li Aristatalis* (Aristotle's Book on Stones), which J. Ruska translated in 1912 and attributed to the date above.[17] It is also worth mentioning that the words *Djawhar* and *Djawahir* refer both to unmounted precious stones and jewelry incorporating precious stones. In North Africa, the word *Djawhar* is also used to describe pearls.

Notes

1. Al Tifaschi, *Azhar al Afkar fi Djawahir al Ahdjar*, 2.
2. Bauer, *Precious Stones*, 103.
3. Al Tifaschi, 2.
4. Webster, *Gems: Their Sources, Descriptions, and Identification*, 242.
5. Ibid., 209.
6. Schumann, *Gemstones of the World*, 100.
7. Al Biruni, *Al Djamahir fi Ma'arifat al Djawahir*, 82.
8. Schumann, 86.
9. Caley and Richards, *Theophrastus on Stones*, 147.
10. Pliny, *Natural History, vol 6, book 37, 464.*
11. Ibn al Akfani, *Nukhab al Zakha'ir fi Ahwal al Djawahir*, 49.
12. Mullet, "Essai sur la Minéralogie Arabe," *Journal Asiatique*, 7.

13. Al Tifaschi, *Azhar al Afkar fi Djawahir al Ahdjar*, 2.
14. Al Biruni, *Al Djamahir fi Ma'arifat al Djawahir*, 210.
15. Bruton, *Diamonds*, 9.
16. *Encyclopedia of Islam*, 250.
17. Ibid., 250.

CHAPTER 3

Weights and Measures Used in the Thirteenth Century

1 *Mna* (also called *Mana* or *Mann*) = approximately 2 *Ratls* = 794 g.[1] The *Ratl* is a measure of weight that varies from country to country. In Egypt, it is equal to 449.28 g, in Syria 3.202 kg, in Lebanon 2.566 kg, and in North Africa (Morocco) 16 oz. Therefore 1 Egyptian *Ratl* = 12 oz.[2]
1 *Mithqal* = 24 carats = 1 3/7 *Dirhem* = 3.12 g. = 0.12 oz.[3]
Qirat = carat = *Tassouj* (1/8th of 1/3rd of 1 *Mithqal*)[4]
1 *Dirhem* = 16 carats; 1/2 *Dirhem* = 8 carats = 2.928 g.
The *Qintar* is a measure of weight that varies from country to country. In Egypt, it is worth 44.93 kg, in North Africa 53.9 kg, and in Syria 256.4 kg.[5] 1 oz = 26 *Dirhems.*
It is assumed that unless otherwise specified, al Tifaschi meant the Egyptian units of weight when he used the words *Ratl, Mithqal,* and *Qintar.*

In addition to the above, seeds were used to weigh small items; thus mustard seeds, grains of barley, pepper seeds, etc., were used to weigh gemstones. *Fulful,* also called *Filfila* (plural *Falafel)* referred to pepper corns, and was used for the measure of small stones. The carob seed was also used to weigh pearls and gemstones, due to the uniformity of its weight, equal to almost exactly 1 carat. However, it was not mentioned by al Tifaschi.

The *Farsakh* (parasang) was used to measure distances. It is now officially estimated at 6 kms.

Numismatics of the Thirteenth Century

DINAR*:* Was the gold unit of currency in early Islam. Al Qalqashandi estimated 1 dinar to be worth 12 *Naqara Dirhems* in the eighth century.[6] The value of the dinar fluctuated between 12.5 and 17.5 *Dirhems* over the different periods. M. Bauer estimated 1 dinar to be worth 3 Pounds Sterling in 1896.[7] Clément Mullet estimated 1 dinar to be worth on average French Francs 14.586 in 1868.[8]

When he speaks of "dinar," al Tifaschi probably meant the Egyptian gold dinar. The word gold *Mithqal* was also used to describe the gold dinar indiscriminately.

Dirhem: Was the silver unit of the Arab monetary system from the rise of Islam to the Mongol period. The value of the *Dirhem* fluctuated over the years, and it is difficult to give an exact estimate of equivalence to the dinar for specific periods. Quoting Vasquez-Queipo, Clément Mullet estimated that the average value of the silver *Dirhem* during the "*Khulafa*" period (period of the various Caliphs), was 0.626 French Francs in 1868.[9]

The Egyptian *Wirq* (also called *Waraq, Wiraq,* or *Warq) Dirhem* was a silver *Dirhem* legally worth 1/3 of a *Naqara Dirhem* (according to al Qalqashandi).[10]

In Yemen, the *Ya'qubi* and *Maliki* dinars were in use. 1 *Ya'qubi* dinar was worth 5 *Maliki* dinars or 1 1/4 Egyptian dinars. 1 Egyptian gold dinar was worth 41/2 *Maliki* Dinars.

In North Africa (Morocco, Tunisia, Algeria, and Libya), the *Maghribi* dinar was used.

1 *Maghribi* dinar was worth 10 *Maghribi Sakkiyya Dirhems.*

1 *Maghribi* dinar was worth 5 Egyptian *Naqara Dirhems.*

1 Egyptian dinar was worth 2.5 to 3.5 *Maghribi* dinars.[11]

Notes

1. Hassan and Khafaga, commentary on al Tifaschi's book, 226.
2. Ibid., 225.
3. Ibid., 210-211.
4. *The Hans Wehr Dictionary of Modern Written Arabic*, 804.
5. Ibid., 793.
6. Hassan and Khafaga, commentary on al Tifaschi's Book, 231.
7. Bauer, *Precious Stones*, vol. 1, 257.
8. Mullet, "Essai sur la Minéralogie Arabe," 504.
9. Ibid., 508.
10. Hassan and Khafaga, commentary, 239.
11. Ibid., 231-232.

PART II
AZHAR AL AFKAR FI DJAWAHIR AL AHDJAR

Different Versions of al Tifaschi's Manuscript: *Azhar al Afkar fi Djawahir al Ahdjar*

THE Bibliothèque Nationale in Paris, France has eight different versions of al Tifaschi's manuscript on gemstones:

Manuscript	No. of pages	Size	Date
1- Arabe 2770	63	17.5cm x13cm	1423
2- Arabe 2773	83	18.5 x 14cm 13 lines/page	13th c.
3- Arabe 2774	67	17cm x 13cm 15 lines/page	15th c.
4- Arabe 2775	74	17.5cm x 13cm 13 lines/page	16th c.
5- Arabe 2776	24	22cm x 16cm 25 lines/page	16th c.
6- Arabe 2777	42	26.5cm x 14.5 cm 25 lines/page (incomplete)	18th c.
7-Arabe 2778		18cm x 13cm 11 lines/page (incomplete)	16th c.
8- Arabe 5291	33	17.5cm x 12.5cm (incomplete)	18th c. (?)

Title	Arabic Title
(*Kitab al Ahdjar*) or Book on Stones	كتاب الاحجار
(*Kitab Djawahir al Ahdjar*) or Book on the Best of Stones	كتاب جواهر الاحجار
(*Kitab Riyadat al Afkar fi Khawass al Djawahir wal Ahdjar*) or Book on the Exercise of the Mind with regards to the Properties of Jewels and Gemstones	كتاب رياضة الافكار في خواص الجواهر والاحجار
(*Kitab al Lla'ali al Mudi'a fi Khawass al Djawahir wal Ahdjar al Mulukiyya*) or Book of Brilliant Pearls on the Properties of Jewels and Royal Stones	كتاب اللآلئ المضية في خواص الجواهر والاحجار الملوكية
(*Kitab al Ahdjar*) or Book on Stones	كتاب الاحجار
(*Kitab al Ahdjar*) or Book on Stones	كتاب الاحجار
(*Azhar al Afkar fi Djawahir al Ahdjar*) or Best Thoughts on the Best of Stones	أزهار الافكار في جواهر الاحجار
(*Al Ahdjar al Mulukiyya*) or The Royal Stones	كتاب الاحجار الملوكية

There are six different versions in the Egyptian Organisation for Books (*Dar al Kutub al Masriyya*), and one in the Arab League headquarters in Cairo, Egypt:

Manuscript	No. of pages	Date
9- No. 136	187	A.H. 697
10- No. 91	101	undated
11- No. 349	132	undated
12- No. 642	74	undated
13- No. 461		
14-	80, 12.5cm x17cm	19th c.

Photocopied version of the manuscript in two parts, B/W. *Azhar al Afkar fi Djawahir al Ahdjar* or Best Thoughts on the Best of Stones.

كتاب أزهار الافكار في جواهر الاحجار

Original manuscript but incomplete; half the manuscript comprises part of a book by al Kindi on gemstones. *Kitab al Ahdjar al Mulukiyya* or The Book of Royal Stones.

كتاب الاحجار الملوكية

Photocopied version of the manuscript in one part, B/W. The following caption is written in English on the cover: "Bought in Cairo in 1822. Presented to the British Museum March 31, 1857 by Major General T. Perronet Thompson, some time Interpreter to the British force in the Persian Gulph (1820); on which occasion (on the 8th of January 1820) was concluded a Treaty with the Arabs being the first public Act in which the Slave Trade was (written?) down Piracy." *Haza'l Kitab yashtamil 'ala Khawass al Ahdjar wa Manafi'iha wa Qimatiha* or This Book is about the Properties, Benefits, and Value of Gemstones.

هذا الكتاب يشتمل على خواص الاحجار ومنافعها وقيمتها

Photocopied version in one part, B/W, from original in the Topkapi Saray, Istanbul, Turkey. *Azhar al Afkar fi Djawahir al Ahdjar*. Under the peridot chapter, it was mentioned that the book was completed in A.H. 640 or A.D. 1240.

أزهار الافكار في جواهر الاحجار

Photocopy of version No. 9 above, but in one part only.

Photos on microfilm from a copy in the Institute of Arabic Manuscripts in the Arab League headquarters in Cairo, Egypt, from an original manuscript in Sheikh Khalil al Khalidy's library in Jerusalem. *Kitab Ahdjar* or Book on Stones.

كتاب الاحجار

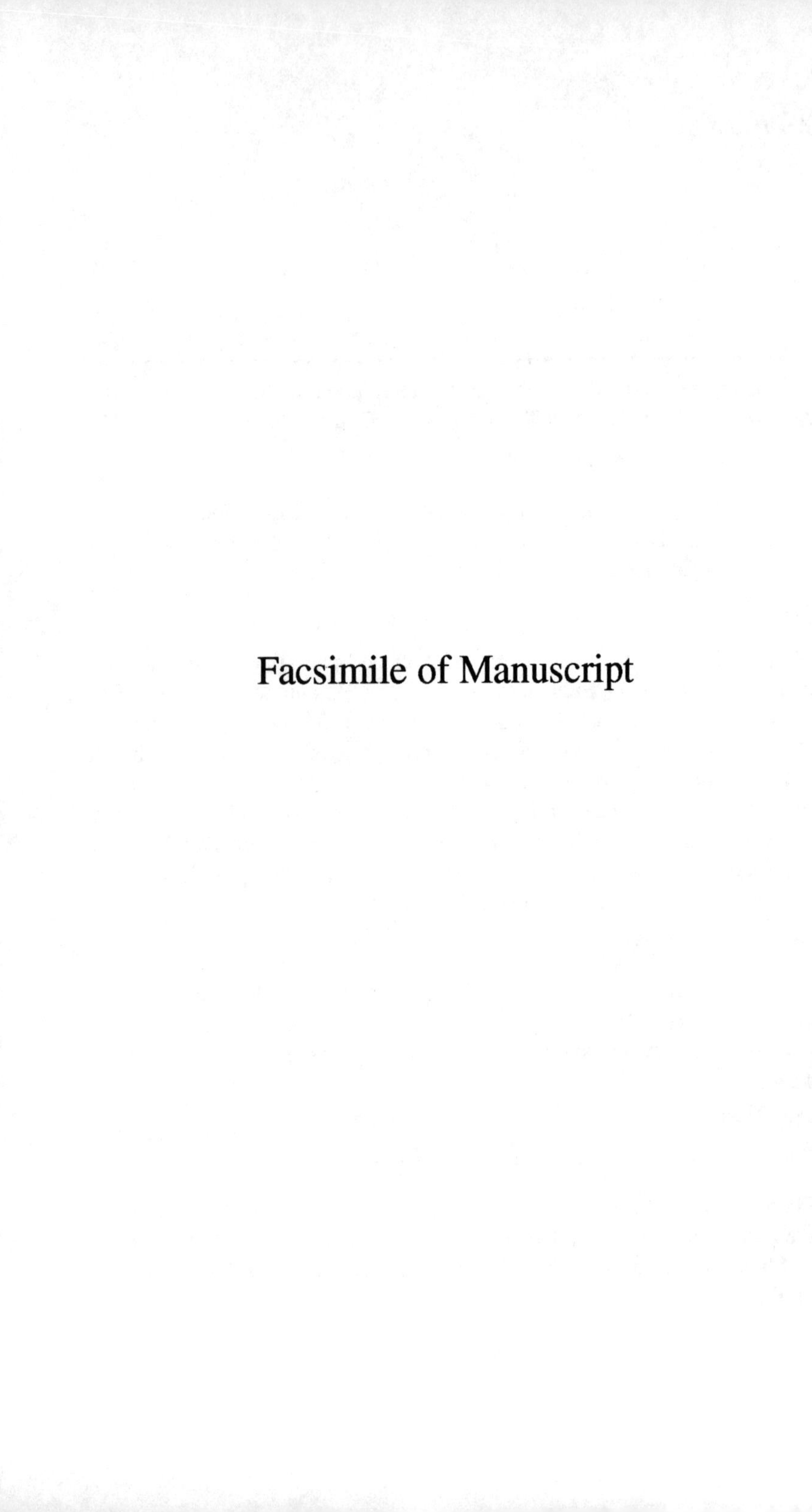

Facsimile of Manuscript

84

التجربة ثم اختلط طريقتها الى الوجود والدال عليها كيف ما
كانت مشكور على بيانه وعند جميع العقلاء محمود
وهذا آخر ما اوردت ارادته والشكر لواجب
الوجود وصلواته وسلامه على انبيائه اجمعين
فالحمد لله على انعامه وجميل عادته وصلى الله على محمد وسلم

تم الكتاب بحمد الله وعونه وحوله وقوته
ولا حول ولا قوة الا بالله العلي العظيم

وصلى الله على خير خلقه محمد النبي الامي وعلى آله وسلم تسليما

بسم الله الرحمن الرحيم
لا اله الا الله
محمد رسول الله

82

فاذا اصابت في الارض تحجر بعضه على بعض طبقة على
طبقة واصله من رطوبة غليظة مائية غلبت
عليها الارضية واليبس وتكررت اجزاؤها واشتد تداخل
بعضها في بعض ولم تكن فيها دهنية كدهنية الذائبة
ولم يقفوا عليها اليبس لكون اصلها من الماء فصارت
لذلك لا تذوب بالنار كما تذوب الاجساد الذائبة
معدنه الذي يتكون فيه الطلق تكون
بجزيرة قبرس كثيرا واما اجلبه جيدة ويكون بجبال كثيرة
غيرها وذكر لي ان نوعا منها معدنيا ومنها الجلب
جبلي وزديه الطلق نوعان فضي وذهبي والفضي
ابيض صافي اللون والذهبي الى الصفرة وهو اجوده
وهو بارد يابس خواصه في ذاته منها
انه اذا دخل النار لم يحترق ولم يتكلس ولم يذب
كالاجساد

كتابه الاحجار ومن هاهنا يقول الحكماء انه اذا
حل وطليت به الاجسام حجبها ان تحرقها النار
ومنها لو دق الداق في هاون الصفر او دقه بمطرقة رق
الحديد وبكل شيء يدق به الاجسام لم يعمل فيه شيء
وليس يحتال عليه بحيلة تسحقه الا بان تاخذ منه
ما شئت وتسحقه وتجعله في مسح شعر اسود او ثوب
خشن مع حصيات ثم تضع الثوب في ماء حار وتدلك
فيه قوما ثم يتخلل جسمه اولا فاولا حتى يستحيل
كله فيخرج ويجمع كالدقيق المطحون فتستعمله فيما شئت
خواصه في منافعه منها انه يفتت الحصاة
في الكلى وينفع المثانة اذا شرب منحالته المستخرجة
على الوجه المذكور آنفا ومنها ما ذكره المسيحي في
كتابه الملقب بالمائة ان الطلق يستعمل في قطع

83

الدم من خارج ولا يجوز استعمالها من داخل البتة
ومنها ما ذكره ابن ضنهار بخت من انه ينفع من الاورام
الحارة التي تكون في الثديين والبيضتين والاربيتين
عند ابتدائها ومن نزف الدم من الرحم والمقعدة
ومن القرحة التي تكون في الامعاء وينفع من نفث
الدم اذا اخذ بماء لسان الحمل وبقاوم السموم اذا
اخذ مع الشراب قال وهو بارد في الاولى يابس
في الثانية وقد ذكرنا عن المسيحي انه قال لا
يجوز استعماله الا بالتعليق من خارج لا غير فيما
ذكره القدماء في استعمال الطلق في حجب الاجسام
عن النار نقلت من كتاب اسرار الخلفاء
للمسعودي في صفة الاطلية التي تكون على السلاح
والخيل فتمنع فيها النار ولا تحترق يوخذ من الطلق
والصمغ

والصمغ العربي من الواحد رطل ومن الاخر اربعة
ارطال وجبس رطلان ومن الدقيق الحواري ما شئت
ومن بزر قطونا لكل عشرة ارطال من الجميع رطل
فيسحق الطلق على الصفة المذكورة فيما سلف ويجعل
مع الصمغ العربي ويخلط بالجبس والدقيق ولعاب البزر قطونا
وتاخذ خل خمر ممزوجه بالماء حتى تنكسر حموضته
ويخلط بالبزرقطونا الملعبه وتعجن الادوية عجنا يمكن
طليه على ما ذكرنا وتدهن به ما شئت فانه جيد
قال ولو طليت به جنة وطرحتها في النار لم تحترق
قال مصنف الكتاب وعلمنا الطلق
طرق كثيرة غير هذه بالتجربة الا انها لا تدخل في هذه
الاعمال التي تحجب النار وهذه الصفة المذكورة لم
اجربها فنقلتها على ما وجدتها عليه حتى تجربها

80

حبتُهُ وَزَردِيَّته اجودُ البلّور اصفاهُ وانقاهُ وابيضُه
واشفُّه واسلمه من التشعير فان كان مع ذلك كبير
الجرم انيةً كان او غير انيةٍ كان الغاية في نوعه
قال يعقوب بن اسحق الكندي ان في البلور
قطعًا يخرج في القطعة منها من المعين اكثر من مائة
من قال مصنف الكتاب وقد اهدى بعض
تجار الافرنجه الى ملك الغرب في عصرنا هذا قبّة من
البلور مصنوعة من قطعتين يجلس فيها اربعة وصنع
ببلاد الافرنجه من اوانيه غرائب رايت عند بعض
ملوك اذربيجه صورة ديك من البلور اهداه له بعض
الافرنجه كمل اربعة ارطال شرابا الا يخل من صورة
الديك ولا الجرم بشيء حتى اظفاره وجميعه مجوّف وشاهد
اظفاره اذا صب فيه الشراب دخل في الاظفار واجتمع
في عنق

في عنق هذه الصورة وسخ سندا اكثر عنقها فطلبت من
يزيله فلم يقدر عليه للخطر المرتكب في ازالته فطلب
احد الخراطين بخمس وطلب خمسين دينارا مغربية على
ازالته والتزم ذلك فتلطفت به واحسنت اليه حتى رضي
واخذ وازال ما كان في عنقه بحيث لم يطلع عليه احد
وامرحه كان لم يكن فيه شي فلاطفت هذا الصانع
المذكور حتى اطلعني على كيفية عمله في ذلك وذلك انه
اراني ذكر ضب اتخذه لهذا الشان ومن شان ذكر الضب
يلتوي بالرطوبة في اي موضع ادخل فادخله بالمخرطه
في عنق الصوره وكان يديره بالمخرطه فيدور على النواحي
وهو يعمل في قلع الوسخ الراسخ في عنق الصوره حتى لم يبق
منه شي البتة واخبرني غير واحد من العجمان بالقرب
من غزنة بينها وبين كاشغر على مسيرة ثلاثة عشر يوما

81

من غزنة واد بين جبلين على ذلك الوادي طريق مؤدٍ
الى كاشغر والجبلان اللذان على الوادي من جهتيه
كلاهما بلور خالص تقطع حجارته في الليل لان اشعته
اذا اطلعت عليه الشمس منع العمل فيه بالنهار ويصنع منه
خوابي الماء في كاشغر وفي غزنة واخبرني بعض اهل
غزنة انه راى في قصر ملك غزنة شهاب الدين الغزنوي
اربع خوابٍ للماء كل خابية تحمل راوية من الماء
من روايا البغال وثلاثة جميعها على محامل يصعد منها
الى الخوابي بدرج والخوابي ومحاملها من البلور
خواصه في نفسه ما ذكر ارسطيس في كتابه
في الاحجار من ان البلور يذوب كما يذوب الزجاج ويقبل
الصبغ قال احمد هذا صحيح الا ان ذلك ليس
البلور من ذاته بل بما يدخل عليه فيوجب له ذلك والا

فالبلور اذا دخل النار سادجا تفتت وتكلس قال
واخبرنا من دخل كاشغر ان خوابيهم للماء من الجبلين
المذكورين القريبين منهم ومنها انه يستقبل به الشمس
ثم ينظر الى موضع الشعاع الذي يظهر من الحجر فيستقبل
خرقة سوداء فانها تحترق وتوقد منها النار
خواصه في منافعه من علقه عليه لم ير منام سوء
قيمته وثمنه البلور يختلف ثمنه بحسب كبر
جرمه وصغره في اوانيه وحجارته فان الغبطة فيه انما
هي بحسب كبر الاواني والالات المصنوعة منه
بالجملة فالانية التي تحمل رطلا اذا كانت صافية سالمة
من التشعير تساوي ثلاثة دنانير مصرية ونحو ذلك

الباب الخامس والعشرون في الطلق

علة تكوّنه في معدنه الطلق يتبع من الهواء كالتلك

78

القلعة الى سائر الجهات البعيدة عنها ومن خواصه
انه يقطع كثرة الاحتلام عن لابسه ۝ ومن خواصه
ما ذكره جالينوس في الادوية المفردة انه ينفع من
وجع المعدة بالتعليق عليها من خارج ۝ قيمته
وثمنه ۝ هذا الحجر يصنع منه اواني تجلب من الصين
ووقفت على صحن منه بيع بسوق القاهرة بخمسة دنانير
مصرية ووقفت على صحن اخر كبير سام صاحبه فيه
بخمسين دينارا فاما خاتم منه فانه يساوي اربعة دراهم
نقرة ويصنع بالهند وان تباع على الهايشم وليس منه
الباب الثالث والعشرون في النصب ۝
علة تكونه في معدنه تقدم القول فيه معدنه الذي
يتكون فيه باليمن ومنه يجلب الى البلاد جيده وردية
منه ابيض وزيتوني والزيتوني اجودهما وفيه نوع

ازرق وهو مصنوع خاصيته ومنفعته قال المسيحي
حجر النصب ينفع المعدة من جميع عللها مخصوصة به
بالتعليق من خارج قال احمد هكذا نقل عن المسيحي
وانا اراه نقلا غير صحيح وان الذي اراد انما هو اليشم
المتقدم المذكور لا النصب والله اعلم ومن خاصية
هذا الحجر في نفسه انه يقبل الصبغ والجوهريون يصبغونه
يشبهون به الورد واكثرها يوجد يكون منه قطعا
صغارا قيمته وثمنه فص منه درهم ونصف بدرهم
الباب الرابع والعشرون في البلور ۝
علة تكونه في معدنه قال بليناس ان البلور
حجر ابو زفي ابيض للاعراض الذي عرضت فيه واصله
الياقوت كما ان الفضة يوقيه التلون واصلها الذهب
كذلك البلور اقول ان المعدن الذي كان فيه

79

البلور كان فيه رطوبة ممزوجة فلما اصابها حر المعدن
كانت الرطوبة عالية على اليبس قاهرة فلما اصابت
الرطوبة حر الشمس سخنت وتغلغلت ودخلت في جسد
اليبس فحللته بلين التدبير وطول المدة فلما انحل صار
اليبس في الرطوبة مائعا صافيا لقهر الرطوبة لها واعتدال
الطبخ عليه فلما ظهر اليبس عليه اجمد ما ابيض معقدا
فصار حجرا ابيض صافيا وانما اصعب على الحمرة رطوبة
واعتدال الحر عليه في معدنه فابيض ظاهره وصار
باطنه احمر وانما تفقع البلور في النار من اجل ملحيته
وانما تولد هذا الملح من قلة الدهن في ظاهره لموضع
البرودة الظاهرة في اعلاه فظهر معها الملح وبطنت
الدهنية في باطنه مع الحرارة فصار دهنه من الملح
واذا اصابه حر النار تفقع ذلك الملح يتفتت جسده

وانما صار الحديد يقع عليه لان رطوبته كثيرة قد رطبت يبسه
فصار رخوا ضعيفا وانما صار صافيا لقلة تكاثف اجزائه
وانما لم تتكاثف اجزاؤه لقلة افراط اليبس عليه وقلة
معونة الحرارة له في معدنه فلم تتداخل اجزاؤه بعضها
في بعض ۝ بعد علة تكون البلور ۝ معدنه
الذي يتكون فيه من البلور ما يتكون بجزيرة العرب
بالحجاز وهو اجوده ومنه ما يؤتى به من الصين وهو
دون العربي ومنه ما يكون في بلاد الافرنجة وهو
جيد ايضا ومنه معادن بناحية ارمينية يميل لونه
الى الصفرة يعرف بالزجاجي كانه مطبوخ بالنار وقد
ظهر في هذا التاريخ معدن بالمغرب الاقصى بمقربة من مراكش
حاضرة المغرب نقي اللون الا ان فيه تشعيرا وكثر عندهم
حتى فرش ملك الغرب مجلسا كبيرا وصار حيطانا ۝

76

اربعة انواع ؛ اوّلها وهو اجودها ما اشتدت ورديته
وسماويته معًا وهو اثمنه ويليه ما اشتدت ورديته
وضعفت سماويته ويليه ما اشتدت سماويته وضعفت
ورديته ويليه وهو ادونه وارداه واقله ثمنا ما ضعفت
سماويته ووردية معًا خواصه ومنافعه الجمشت
حجر كانت العرب تستحسنه وتزين به الاتها واسلحتها
وعلاجه في قطعه وجلاه كعلاج الزمرذ اعني انه
يجلى اولا بالسنباذج على لحت الاسرب بالماء ثم يجلى
بعد ذلك على خشب الغرب فمن خواصه ما تذكر العرب
من انه يشجع لابسه في الحرب ومنها ما ذكره الرازي
في كتابه الذي سماه تحفة الملوك في ان من خاصية
حجر الجمشت ان من صنع منه قدحا ثم شرب به ما شاء
به ما شا من النبيذ لم يسكر ومنها ان لابسه آمن من

حدوث

حدوث النقرس ومنها انه من وضعه تحت وسادته
أمن من الاحلام السوء قيمته وثمنه فص زنة
نصف درهم منه بدرهم ؛ الباب
الحادي والعشرون في الخماهان ؛
وهو المسمى حجر الحديد علة تكونه في معدنه هذا
الحجر اسود حديدي وقد تقدم القول في علة تكون
الاحجار الحديدية ؛ معدنه الذي يتكون فيه هذا
يجلب من الكرك على مسيرة سبعة ايام من مصر ومنه
يجلب الى ساير البلاد جيده ورديه اجوده الاسود
الشديد السواد الذي يضرب الى الحمرة الحديدية ؛
خواصه ومنافعه خاصية هذا الحجر النفع من سكر
الشراب الصرف بذلك عرف ومنه سمي وذلك بان يحك
ويشرب حكاكته له في ذلك اثر ربحي ومنفعة ظاهرة

77

قيمته وثمنه زنة رطل منه بمصر ثلاثة دراهم نقره
وهو بغير مصر اعلا منه بمصر لقرب معدنه من مصر ؛
الباب الثاني والعشرون في اليشم
علة تكونه في معدنه اليشم واليشب وهو ايضا اليصب
حجران فضيان كانهما قريبان من بعضهما بعضا ويكونان
في معادن الفضة من ابخرة مقصرة عن كيان الفضة
بالزيادة والنقص في الكيفيات الاربعة حسبما بيناه
فيما سلف معدنه الذي يتكون فيه كاشغر ومنه
يجلب الى ساير البلاد وكاشغر اقليم منه مدينة كبرى
بين الصين وبين مدينة غزنة على نيف وعشرين يوما
من غزنة الى جهة الشمال لسانهم تركي جيده ورديه
اليشم نوعان احدهما ابيض والاخر اصفر كلون العاج
العتيق وهذا هو الجيد الخالص منه المعدني ؛ فاما

الابيض فانه مصنوع يصنع بالصين من اخلاط مجموعة
وليس فيه شي من خواص اليشم ومنافعه وانما هو يشبهه
لا غير وصنعت انا بالقاهرة المعزية كلاها الله من
هذا اليشم اواني واهديتها لبعض الامرا ممن يقتني اليشم
ويحرص عليه وعنده منه اواني ولم يشك ان ما اهديت
له من معمول الصين فعرفته اني عملته فانكر ذلك
حتى اوقفته على الدليل فيه تصنعت له اواني على
شكل مخصوص فصدق عند ذلك ؛ خواصه
ومنافعه خاصية هذا الحجر ان الصاعقة لا تنزل على
موضع يكون فيه اخبرني ثقات من العجم انهم
شاهدوا قلاعا ببلاد الترك حيث تقع الصواعق
كثيرا فتبني في القلعة منارة ويعلق هذا الحجر فيها
ظاهرا فترى الصواعق نازلة من السماء تحيد عن

74

اذا شرب نفع من نفث الدم ومن الشحج نفعا بينا ومنها
اذا شرب نفع من عسر البول ومنها اذا شرب بالماء
حلل ورم الطحال ومنها اذا علق على المعدة نفع
من جميع عللها منفعة عظيمة حتى يقال انه يفعل في ذلك
كفعل الزمرد الذبابي ومنها انه يقطع نفث الدم من الجسد
كله وصفة شربه لنفث الدم ونزفه ان يسحق بعد الحرق
الذي بيناه فيما سلف يوخذ منه ثلاثة دراهم مع دانق
ونصف صمغا عربيا ويسحق ببياض البيض ويشرب بماء
بارد فانه ينفع من النفث والنزف نفعا عجيبا ومنها
ما ذكره الاسكندر من ان المرجان اذا علق على رجل
من به النقرس نفعه قيمته وثمنه قيمة المرجان
باوقيه وبها معدنه كما بيناه فيما سلف الرطل المغربي
من خمسة دنانير الى سبعة من الدنانير السكية المغربية

التي

التي دينارها عشرة دراهم سكة باصطلاحهم وهي
خمسة دراهم نقرة وذلك الدق منه الخام غير المثقوب
ولا المصنوع فانه انما يجلى ويثقب ويصنع بالاسكندرية
فيكون على ضعفي ما ذكرناه من سعره باوقيه وثلاثة
اضعافه على قدر كثرة جلبه وقلته وكبار اغصانه اكثر
ثمنا من صغاره الباب التاسع عشر في
السبج علة تكونه في معدنه السبج من الاحجار
الرصاصية وقد تقدم القول في ان كل حجر ينسب الى احد
الاحجار الذائبة فانه ابتدا في اصل تكونه ان يكون حجرا
منها فانعقد عن ذلك لبعض الاعراض الداخلة عليه
من زيادة حر او برد او رطوبة او يبس ونقصانها حسبما
بيناه فيما سلف معدنه الذي يكون فيه
السبج يوتى به من موضعين احدهما الهند والاخر

75

بلد فارس جيده ورديه اجوده الهندي وهو
حجر اسود شديد السواد ليس فيه شفوف سوى
انه يرى الوجه كالمرآة براق رخو شديد الرخاوة
سريع الانكسار خواصه ومنافعه السبج بارد
في الاولى يابس فيها وهو نافع في اكحال العين
وقال ارسطو من خاصية حجر السبج ان الانسان
اذا اصابه ضعف في بصره عند الكبر عن علة حادثة
تعسر عليه ان ينظر الى شي حتى يرى خيالا كالغمام
او كالذباب او كالضباب وكل هذا يدل على ابتدا
نزول الماء في العين ثم اتخذ من السبج مثل المراة
وادمن النظر فيها امسك البصر وقواه وشده ودفع
عنه العلة النازلة به ومن خواصه انه ان عمل
منه فص خاتم ولبس وادمن النظر اليه احد البصر

ومن خواصه انه من علق عليه منه خرزة حرزه
من الناس لبعض ثوابهم او جعله فصا لخاتمه دفع عنه
ابصار الذين ينظرون اليه باعين رديه وقال
ابن ضهار بخت انه ينفع العين اذا سحق وادخل في
الاكحال ويقوي نورها وان اكتحل منه مع بزر الكلثم
ومرارة باشق ومسك نفع ذلك قيمته وثمنه
خرزة مثقال بنصف نقرة الباب العشرون
في الجمشت علة تكونه في معدنه هذا الحجر
حديدي ابتدا في معدنه ليكون حديدا فانعقدته
الاعراض الداخلة عليه بزيادة الطبايع ونقصانها
معدنه الذي يكون فيه يوخذ الجمشت بقرية تسمى
الصفرا مسيرة ثلثة ايام من مدينة سيدنا رسول
الله صلى الله عليه وسلم جيده ورديه الجمشت

72

ويديرون الشبكة حول الشجرة حتى تلتف بها ثم يجذبونها
حتى تنقلع وتخرج من الماء ابيض لينا فاذا اصابه الهواء
جف واحمر وتزال اصوله ناحية فهي البسذ وتفصل اغصانه
قطعا صغارا وكبارا على قدر العقد والشعب التي
تكون فيه فهي المرجان فيجاء عند ذلك على مسن بالماء
ويجلى بالسنباذج ويثقب ان اريد ثقبه بالحديد الفولاذ
المسقى ¤ معدنه ¤ الذي يتكون فيه ¤ المرجان يوجد
في موضع يسمى مرسى الخرز في بحر افريقية ويوجد ايضا
في بحر الافرنجة الا ان الاكثر من مرسى الخرز ومنه يجلب الى
المشرق والى اليمن والهند وسائر البلاد ولا يوجد بغير هذه
المواضع كما يوجد بها منه الكثرة والكبر والجودة ¤
جيده وردية اجوده ما عظم جرمه واستوت قصبته
واشتدت حمرته وسلم من السوس وهي خروق توجد في
باطنه

باطنه حتى يكون منه شيء حاويا كله كالعظم وهي معيبة
وردية وكذلك ما مال منه الى البياض ونقصت حمرته
فهو معيب والعقد والتشطيب من عيوبه الا انها لازمة
له لا تكاد تفارقه لكونه كان اغصانا متشعبة كما ذكرناه
وقل ما يوجد منه قطعة كبيرة مصمتة مستوية لا عقد
فيها ولا تشطيب الا نادرا واذا وقعت وكان لونها شديد
الحمرة كانت النهاية في الجرم وربما كانت منه قطعة كبيرة
مشطبة فيحت حتى تزال تشطيبها وعقدها واملست
واستوت الا انها تنقص بهذا العمل كثيرا ويحسب جودتها
تكون الزيادة في ثمنها ويقلع من المرجان قطع كبار
نادرة ترفع الى ملك افريقية يصنع له منها محابر ونصب
سكاكين ورايت منها محبرة طولها شبر ونصف في
عرض ثلاث اصابع وارتفاع مثلها بغطائها في غاية

73

الجرم وصفاء اللون وحسن الجوهر خواصه في ذاته
منها انه اذا القي في الخل لان وابيض وان ترك فيه
انحل ومن الناس من يتخذ منه فصوص خواتم فاذا اراد
ان يكتب على شيء منها ما احب حمل على جميع الخاتم او الفص
شمعا ثم عمد الى موضع النقش منه فكتب برأس ابرة ما احب
حتى ينكشف الشمع عن موضع الكتابة لا غير ثم القاه في
خل حاذق يوما وليلة او يومين وليلتين ثم رفعه وازال
عنه الشمع فانه يجد موضع الكتابة محفورا قد تأكل
بالخل وبقية الفص على حاله لم يتغير ¤ وقد جربت
ذلك فكان كما ذكرت ¤ ومنها انه اذا القي في الزيت
اظهر حمرته واشرق وحسن فعله ويفعل فيه ضد فعل
الخل والمرجان بارد في الاولى يابس في الثالثة يقبض
ويبرد باعتدال خواصه في منافعه منها ما

ذكره الاسكندر من ان المرجان اذا علق على المصروع
نفعه ¤ ومنها انه يحفظ من الاعين السوء والانفس
الخبيثة الانسية والجنية اذا علق على الاطفال الصغار ¤
ومنها اذا احرق واستن به زاد في بياض الاسنان وجلاه
وقطع الحفر منها وقوى اللثة وصفة احراقه ان يؤخذ
منه شيء ويصير في كوز فخار جديد ويطين رأسه ويوضع
في تنور فيسجر الليل كله ويخرج من الغد ويسحق ويستعمل ¤
ومنها انه ينفع من وجع العين ويذهب بالرطوبة
واذا اكتحل به يقطع اللحم الزائد في قروحها ويجلو آثارها
ويملأ القروح العتيقة لحما وينفع من ظلمة العين وينفع
من ظلمة وجعها وبياضها وكثرة رمصها ¤ ومنها انه
ينفع من الخفقان اذا شرب مسحوقا ويجعل في الادوية
التي تجلو دم القلب الجامد وينفع من ضعفه ¤ ومنها

70

بالماء العذب فانه يجلب بقوى نارية ملطف حتى يذوب
ثانية بيحرك بالانتظام المذكور فان خرج جوهر اللازورد
فهو لازورد عتيق خالص كثير الجوهر سهل الخروج وان لم يخرج
جوهره بهذا العمل والا القي عليه ما يخرجه وهذا موضع
سر في عمله تكلم بابعد بل هو مما يضن به صناعه فان
اللازورد يتلف في هذا الموضع ان لم يعرف هذا السر منه
ولم انقله من كتاب بل هو من جملة ما وقفت عليه
بالتجربة من صحيح كتبنا في الاعمال الصناعية والذي يخرج
جوهر اللازورد اذا تعذر خروجه لها انما هو الزيت
المعتصر من الزيتون او الصابون المعمول من زيت الزيتون
يلقى عليه ايهما حضر فان اللازورد عند ذلك يقذف
صبغه ويخرج جوهره حتى لا يبقى في الارضية منه البتة
فيسكب في اناء نظيف صيني او غضار محكم الدهان

[illegible]

ويترك حتى يرسب جميع ثقله وتذاه وارضيته المختلطة
بجوهره من تراب المعدن وتوخذ ما يطفو على وجهه
من صبغ اللازورد وجوهره الخالص فيرفع ويستعمل هذا
العمل الملتة واقل واكثر على حسب جودة الحجر ورداءته
واحكام الصنعة في اخراج جوهره كما ذكرنا والجهل
والخطا فيه يتلف اكثره او جميعه م

الباب الثاني عشر في المرجان

علة تكونه في معدنه يكون المرجان متوسطا بين حالتي
الجماد والنبات وذلك انه يشبه الجماد بتحجره ويشبه
النبات بكونه اشجارا نابتة في قعر البحر ذوات عروق
واغصان خضر متشعبة ما تثمر ثم يلين والعلة
في ذلك امتزاج الحرارة بالرطوبة في قعر المعدن وعليه
الرطوبة على الحرارة لمجاورة الماء فالمرجان يشبه المعدن

71

بحسبك ويشبه النبات بوجه وذلك انه لما طال مكثه
في الارض وافرط في حر به ونحى ذلك الماء بما يصل اليه
من حر الشمس وتلطف وقوى على تحليل يبس الارض بلينه م
والحر الذي فيه بما اقتبسه من الحر الناري فلما انحلت
اجزاوه بلين الماء وصارت منحة لينة ونخت عليه
الشمس بحرها قوى على نسف الحجر واليبس من الشمس لما فيه
من يبس الارض المتحن في بطنه فلما تكاملت اجزاء اليبس
فيه بافراطه وحركته الحرارة وهو رطب طلع الى الهواء
فلما اصابه الهواء ذهبت الحرارة منه فبرد وجمد وصار
حجرا في الهواء جامدا والطبايع ذاتية تعمل فكلما هرب
ذلك الماء من حرارة النار دفع مما يليه وتصاعد طوالع
في الهواء وظهر جزء النار حتى استطال وصار نباتا عاليا
ابيض وانما احمر لذهاب الرطوبة عنه وظهور اليبس

فلما اشتدت عليه حرارة الشمس ذهبت الرطوبة من ظاهره
فكانت الحمرة من شدة الحرارة فكان احمر فهذه علة
حمرته وانما قلنا انه نباتي لطلوعه في الهواء وتشعبه
كاغصان النبات وانما صارت له اغصان لان الماء
الممزوج باليبس لما اصابه حر النار هرب منها فوقا
في الهواء متدافعا فلما طلع الى الهواء تشعب على ما كان
من حر النار وقطع مقدار قوته فهذه علة تشعب اغصانه
وهذه علة السبب في تكونه في معدنه قال احمد
وكذلك يكون في الماء حيوانات من عالمي الحيوان والنبات
كالاسفنج البحري فانه ينبت على الاحجار ملازما لمكانه
فاذا لمس واحس باللمس انقبض ولين له حركة استقالية
كالحيوان وبحر المرجان غاصة يتخذون لبنا كاونة
من قضب مثقلة بالرصاص يلقونها على حجر المرجان

68

فان نفع نفعاً عجيباً من حمى الربع الا انه ينقص لموس الحر
السودا انقصاً معتدلاً لا يعدله شي من الادويه وينفع
ايضا اصحاب الربو ومنها انه يذهب الطمث وينفع من وجع
المثانه ويقلع الثاليل واذا سحق على شحيره نحيل
للقرحه الذي تاكل اللحم ويخرج الحسد واذا اطلي مسحوقا
بالخل على البرص براه فان كان مغسولا اخرج السودا
بلا قي من غير ان يهيج في البدن حرارة وان شرب غير مغسول
اخرجها بالقي ومنها انه ان سقي بما العسل نفع من وجع
الكبد ومنها ان سحق بالخل وطلي على البرص نفعه
ومنها ان علق على صبي لم يفزع ومنها انه يجعد الشعر
ويحسنه اذا جعل في بعض ادهان الشعر اي دهن
كان دهن به وطبع اللازورد البرد واليبوسة كما
المسيحي في كتابه المعروف بالمائه حجر اللازورد يسهل

السودا بقوة ولا يعقب ضررا ولا كربا الا انه بطي عسر
الفعل ينبغي ان يغسل مرارا كثيرة بما عذب ويسحق جدا والا
اورث القي والغثيان ويخلط فيه شي من الافاويه والشربه
الوسطى منه مثقال الى ونصف والحجر الارمني يفعل فعله
ولعسل غسله ويصلح كل صلاح الا ان الشربه الوسطى
منه ثلث مثقال وقد يصنع باللازورد بالكيفيه التي انا
واصفها خصوصا يوخذ جزء من الزرنيخ الاحمر ومثله من
الزرنيخ الاصفر وربع جزء زاج كرمانيا ومثله رمل زجاج
نقيا من التراب خالصا يدق كل واحد على حدته وينخل
ويسقى الخل ثم تاخذ فخارة فتطينها بطين محكم فيه
شعر وسرقين تطيينا جيدا وتتركها حتى يجف ثم يجعل
فيها الادويه وفيها دراق الخل بقدر لبن السويق ويسد
راس الفخار بخرق وتطين راسها ثم سخن التنور بحطب

69

ثم بالسرقين ثم تنشر السرقين فيه بقدر عظم الذراع ثم توضع
الفخارة في التنور وتدفن تحت السرقين ويطين راس التنور
وكوته من اسفل وتترك ليلة فاذا اصبح اخرجت الفخارة
من التنور واخرج ما فيها فانك تجد فيه فصوصا حمرا كالياقوت
مدورة مثل السدر وطوالا احسن ما ترى من الفصوص
ما ثبت لهم اعمل لمن يعمل عليه اللازورد المعدني
ولم ينكر منه شيا وانما ذكرت هذه الصفة لتعلم ان اللازورد
فيه المعدني والمصنوع وهو اقبل شي للغش والدنس ويصنع
على طرق كثيرة غير هذه الا ان هذه اعظم طرق مصنوعه
تدخل في النقوش والاصباغ ولا تكون باشد من المعدني
وامتحان اللازورد الخالص المعدني الذي يراد استعماله
في الادويه يكون بالقايه على الجمر كما بيناه فيما سلف
فان بقيت ولم يستحيل فهو خالص بصمته وثمنه

اللازورد الخالص منه فص ومنه حجر والحجر منه منخول
مغسول مصول ومنه خام والفص الجيد منه الخالص الذي
يصلح للخاتم من ثلثه دراهم نقره الى قريب من ذلك والحجر
المصول المغسول المسحوق من دينار الاوقيه فما دون
ذلك والحجر الخام على ثلثي سعر المغسول فضلا عن الا انه
ينقص في السحق والمصول الثلث مما يقاربه
من غسل اللازورد وتصويله صناعة مربحة يعيش بها
وينتفع من يحكمها منها وصفة ذلك يوخذ المعدني
منه الخالص المختبر بالنار كما ذكرناه فيما سلف
فيصنع منه خميرة وهي راتينج جزء وكندر جزء ويجعل
على النار في مذابة صفر مرتكبة على نار لينة حتى
يذوب ويسحق اللازورد ويعجن بالماء ويلقى في المذابة
ويحرك حتى يختلط الجميع فاستطاع من صفرة ترك ثم يغمر

66

منه صفيحة صحيفة تسعة وبلا بين رطلان ومنها انه
اذا نقع في الزيت استدت خضرته وحسن وان غفل
عنه حتى يطول لبثه في الزيت مال الى السواد ومنها
انه من سقى بحكه او بحله وشفاه انسانا كان
سما ويغط الامعاء ويحدث في البدن سقما الايد هب
ومنها ان من سقي من محكه او نحله ومن امسكه في
فيه ومص ماءه كان له رديا ويجب ان يبادر الى
علاجه بسقي الشراب العتيق ويجعل له في اطعمته الزبد
والسمن ويعالج بما يعالج به شارب الزنجار ومن خواصه
ومنافعه منها انه اذا مسح به على موضع لذع العقرب
سكنه بعض السكون ومنها انه يمنع الحبل شربا الا ان
شربه خطر كما ذكرناه انفا ومنها انه ان سحق منه شي
واذيف بالخل ودلك به موضع القوابي الحادثة من المرة

السودا

السوداء اذهبها ومنها انه ينفع السقيفة في الرأس وفي
جميع البدن ومن الخواص المجربة في الاوبدى منه اذا
حل بالماء وقطر في العين نفع من البياض وازاله في ثلاث
دفعات ومن عجايب خواصه انه اذا سقي من محكه
او سحالته شارب السم نفعه النفع وان شرب منه من
لم يشرب سما كان سما مفرطا يعطب الامعاء ويلهب البدن
ويقتل سريعا ولا سيما اذا حك بحديد وسقي منه بحك
الجسم ويسكي الامعاء ولا يبرا اشاربه ومن خواصه
انه اذا سحق منه شيا والقاه في الزيت الذي يكسر عند
بطريق الصاغة له الحشوبة قيمة لينه واذهب خشنه وكسره
وان خلط مع حجر التنكار كان قوى لفعله في ذلك وقيل
انكرته حجر الذهب ايضا وضوله وهو معتدل في الحر واليبس
وقيل انه حار في الرابعة خواصه انه من قبل به سبع

67

ذبابات ثم دقه وذافه بماء وطلى به ابرا من لذع العقارب
ومن القوابي قيمته وثمنه الاوندك منه العتيق
الخالص الجامع للصفات المذكوره مثقالين المثقال
الباب التاسع عشر في اللازورد
علة تكونه في معدنه تقدم القول فيها آنفا مع الدهنج
معدنه الذي يكون فيه اللازورد يجلب من جبل بطخار
في موضع يسمى حسان من ارض فارس قريب ناحيه ارمينيه
حيث ورد به اللازورد حجر رخو طيني واجوده ما
اشرقا واصفاه لونا السماوي المستوي الصبغ الى
الكحلة ما هو خواصه في ذاته هو منها اذا جمع الى
حجر الذهب ازداد كل منهما حسنا بصاحبه في اعين
الناظرين وان كانا لا يستحيلان عن كيانهما ولا يزيدان
ولا ينقصان في انفسهما الا انه يحسن لون كل واحد

بصاحبه في اعين الناظرين كانهما شكلان متفقان
ومنها انه اذا وضعت منه قطعة في جمر لبس له دخان
خرج لسان من النار منصبغا بصبغ اللازورد على ما هو
عليه وبهذه المحنة يختبر خالص اللازورد ومغشوشه
ومنها ان كلس تكليس ومكث فيه النار حتى اصبه
في منافعه منها انه ينفع العيون اذا جعل في الاكحال
ومنها انه ينبت شعر الاجفان وينقي رطوبة الاخلاط
الحادثة في العيون ويرد العضو الى مزاجه الاصلي الذي
به يكون نبات الاهداب ويزيدها ويقويها ويثبتها
فيها وينبغي ان يسحق حلا ويستعمل كما يستعمل الزرو ز
ومنها ان شرب مغسولا اسهل وان شرب غير مغسول
قيا ومنها ان يمنع الماليخوليا ويسهل المرة السودا
ومنها انه ان شرب منه اربع قراريط بشراب ورد وما

64

في الحجر وان اخذ لم يصل اليه قبل الاسكندر حيلة
وردية السنباذج كانه حسد من الرمل وفيه
حجارة متجسدة صغار وكبار وجوده الكبار النقية
خواصه في منافعه قوة السنباذج البرودة
في الدرجة الثانية واليبس في الثالثة وخاصته انه اذا
شيء كان اكثر عملا منه على جسمه غير مسحوق
وياكل اجسام الحجارة كلها اذا ذلك بها بايسار رطبا
بالماء وفيه جلاء شديد وتنقية للاسنان وله جذبة
يسير ويستعمل في الادوية المحرقة المجففة وله في
الادوية المحللة اثر يحلل الجسد ويغير الاسنان
وان احرق بالنار وسحق والقي على القروح والبثر
في العين الذي قد طال مكثها ابراها ونفع في اخلاط
المراهم المعفنة والمراهم المحرقة وينفع اللثة المسترخية

قيمته وثمنه الاوقية منه بدرهم فيما تقارب ذلك

الباب السادس عشر في الدهنج

علة تكونه في معدنه قال ارسطو ان النحاس في معدنه
اذا الحجر ان تقع له بخار من الكبريت المتولد فيه ويرتفع
ذلك البخار مثل الدخان فاذا صار الى موضع يصمته
الارض تكاثف ذلك البخار بعضه على بعض ثم انعقد
حجرا فكان منه الدهنج قال بلينوس ان الدهنج
واللازورد والشادنه وجميع الاحجار النحاسية انما
ابتدأت في معدنها لتكون نحاسا فلما ابتدأ الزيبق ليتكون
في معادنه وامتزج بالكبريت غلبت الحرارة على الرطوبة
المهياة في المعدن لتكون زيبقا فلما استبدت عليه
الحرارة احمر وصار مثل الشادنه وجميع الاحجار الحمر
فان كان في معدنه شيء من رطوبة انعقد حجرا اخضر

65

لانه شربه من فضله نحاسية غليظة مثل الدهنج فان
افرط عليه يبس الارض زاد سواده مع الخضرة المنبجسة
فيه فصار لازورديا فكان منه حجر اللازورد وغيره
من الاحجار المشيدة النحاسية على قدر الزيادة والنقصان
في الحر والبرد والرطوبة واليبس في تدبير هذه الاحجار
استخرج منها نحاسا الطف من نحاس المعدن وذكر
يعقوب بن اسحق الكندي ان الدهنج اذا سحق بالنطرون
والزيت خرج منه نحاس ناعم احمر اللون ليس بجيد
معدنه الذي يتكون فيه ليس يوجد الدهنج الا في
معادن النحاس والعلة في ذلك ما ذكرناه من اصل تكونه
من الحرارة الا انه لا يوجد كل معدن من معادن
النحاس واكثر ما يوجد في معادن كرمان ومعادن
سجستان من معادن فارس ومنه ما يوتى به من عماز

لبني سليم في برية المغرب وبالجملة فمواضعه كثيرة مختلفة
بحسب اختلاف معادن النحاس الا ان اجود انواعه
اربعة الافرندي والهندي والكرماني والكركي
جيده ورديه اجود الدهنج الاخضر المشبع الخضرة
الشبيه اللون بالزمرد المعروف بخضرة حسنة الذي
فيه اهلة وعيون بعضها من بعض حسان الصلب الاملس
الذي يقبل الصقال الدهنج صفة العتيق منه ولا يكاد
يوجد الا في الافرندي منه لا غير خواصه في ذاته
حجر الدهنج فيه رخاوة ومن خواصه انه اذا صنع منه
آنية او نصب السكاكين ومرت عليه مدة سنين انحل
لرخاوته وذهب نوره ومنها انه اذا حل انحل سريعا
واذا انخرط انخرط جزرا وارابي وغير ذلك وذكر
يعقوب بن اسحق الكندي في كتابه في الاحجار انه راى

62

حتى اصحه ومنافعه ه منها ما نقله احمد بن ابي
خالد المعروف بابن الجزار في كتابه الاحجار عن سلمويه
الطبيب ان حجر الماغنيطس ينفع من النقرس في اليدين
والرجلين اذا امسك في اليد وينفع من الكزاز ومنها
ما ذكر ارسطوان المراة اذا امسكت حجر المغنيطس
في يدها سهلت عليها الولادة ه ومنها انه من شرب
من سحالة الحديد او بعض السموم التي تخالطها الحديد
او جرح بحديد مسموم ثم سحق هذا الحجر وذلك ببعض
الالبان او بماء فاتر وسقي المسموم بالحديد فانه يفرغه
كله من معدته بالقيء حتى لا يبقى منه شيء البته
ويبطل فعل السم ويخلص منه ه ومنها انه ان
سحق ونثر على موضع الجراحة بحديد مسموم ابراه
بخاصيته ونزع ما فيه لذلك ه ومنها انه اذا سحق

وطلي به مع لبن جارية اخرج الازجة والنصول من اللحم
وابرا جراحها ه ومنها انه اذا امسك باليد وعلق
على من يشكو وجع المفاصل من اي نوع كان ذلك ابراه
منه ه وذكر ارسطو ان هذه الاحجار الماغنيطسية
منها ما يلتقط الذهب والفضة والصفر والنحاس والرصاص
والشعر واللحم والاظفار ولـ والحجر الذي يجلب
الذهب فهو حجر اصفر مشرب غبرة شيئا قليلا طبيعة
الحرارة واليبس فان نحل الذهب بما رد الحديد وخلط
بالتراب وامر عليه هذا الحجر اخرجه من التراب حتى
لا يبقى منه شيء البته ه واما الحجر الذي يختلس الفضة
فهو حجر طبيعته البرد والرطوبة وهو ابيض مشرب
غبرة اذا عمل عليه انسان ضر كما يضر الرصاص
وليس فيه شي من الرصاص وليس منه شي في الاحجار

63

يختلس كاختلاس هذا الحجر في قوته الغريزية اذا اخذ
منه زنة اوقية او اقل من ذلك ثم وضع من الفضة على
قدر خمسة اذرع اجتذب الفضة وان كانت مسمرة
قلع ذلك المسمار من موضعه ه واما الحجر الذي يختلس
اللحم فان منه حيوانيا وغير حيواني ه فاما الحيواني
فهو راس ارنب البحر فان ارسطو ذكر ان ارنب البحر
راسه حجر وانه يلتصق باللحم حيث وجد اذا لم يكن
عليه شعر ولا ينقلع عنه حتى يقطعه ثم تقرح موضعه
قرحا رديا لا يكاد يبرا ولا يسيل من ذلك الموضع الذي
يقطع منه اللحم دم ه واما الحجر الاخر فهو غير حيواني
وانه اذا الصق باللحم اقتلعه من لحم الحيوان الذي فيه
روح فاذا الصق بلحم ليس فيه روح نقش الحجر من اللحم
شيئا يسيرا اذا طبخته صار مثل لحم المنخوف هذه

الاحجار التي تلتقط ما وصفناه فيما ذكر ارسطو قال
فاذا كان شيء من هذه الاحجار اي حجر كان منها في
اتانين يوقد فيها النار اتون بعد اتون ثم اضيف اليها
حجارة الكبريت احرق كلما يقرب منها كما تحرق النار ه
قيمة المغناطيس وثمنه اوقية من خالصه
القوي الجذب ربع دينار ه

الباب الخامس عشر في السنبادج

علة تكونه في معدنه تكون السنبادج على نحو ما
تقدم القول فيه من تقدم الماس الا انه دونه بكثير
ونقص عنه في الطبع وكانه نوع منه قصر في كيانه عنه
معدنه الذي يكون فيه السنباذج على نحو ما
تقدم يقال انه يوجد مع الماس ويذكر ان الوادي الذي
يوجد فيه السنباذج به هو باقصى الصين في جزيرة

60

فى الأحجار انه ان لف الجزع بشعر امرأة اضر بها الطلق وعلق عليها ولدت مكانها ٭ ومنها ان وضع الجزع قريبا من النفساء دفع عنها الصفور وخفف جميع اوجاعها ٭ ومنها انه اذا جلى به مسحوقا اصناف اليواقيت حسنها واظهرها نورا واشراقا وبريقا لا يفعله غيره من الاحجار وطبع الجزع البرد واليبس فى الثانية وقيمته خرزة منه مثقال منه بدرهمين نقره ٭

الباب الرابع عشر فى المغناطيس ٭

علة كونه فى معدنه ذكر ارسطو ان الاحجار المغناطيسية كلها ابتدأت فى معدنها لكون حديدا تعرض لها الحر واليبس فصارت حجارة يابسة صلبة شديدة وانما اشتدت هذه الاحجار لشدة الحر فى معادنها وقلة الرطوبة فيها وغلبة اليبس المتصل بها ولذلك صارت حجارة سودا من كان الحديد فهي تجذبه لما بينه وبينها من المناسبة الطبيعية والموافقة والمعاشقة فى اصل الكون حتى انه بلغ من شدة طاعة الحديد له انه ان اخذ قطع حديد رقاق مثل المسال واثبت فى الارض ثم يوصل بواحد منهن الحجر فاذا التصقت قرنتها الى اخرى فتلصقت بالاخرى التي هي ملصقة بالحجر حتى يظن الناس انها منظومة ٭ معدنه الذي يكون فيه ٭ معدن هذا الحجر فى جبل فوق الساحل الذي بين بحر الحجاز واليمن وله ايضا معدن بصنعاء اليمن وذكر ارسطو ان له جبلا فى البحر وان السفن اذا قربت من ذلك الجبل لم يبق فيها شيء من الحديد الا بادر مرتفعا من جروف السفينة نطيرها كما يطير الطير وان كان مسمارا قد سمر لحاجته انقلع حتى يلصق بحجر المغناطيس ولهذا

61

لا تسمر السفن السالكة فى ذلك البحر بالحديد وانما تخرز خرزا بالليف للنارجيل ثم يدسر بخشب لين بقي من الماء واهل اليمن يخرزون سفنهم بقضبان جريد النخل واما جبل المغناطيس فهو فى ساحل بحر الهند وقد قطع منه حجارة عظام كحجارة البنيان وبنوا بها بيتا مرتفعا وجعلوا اسقفه وارضه منها وصنعوا ابدا وهو صنم يعبدونه من حديد واقاموه فى الهواء فى وسط ذلك البيت فبقي معلقا فى الهواء بجذب المغناطيس له وتكافيه فى الجذب من كل جهة من الجهات ٭

جيده ورديه اجود حجارة المغناطيس ما قوى جذبه للحديد وكان لونه لازورديا كثيفا ليس بمفرط الثقل خواصه فى ذاته ذكر ارسطو ان حجر المغناطيس ان انقع فى ماء الثوم او البصل وترك فيه ثلثة ايام بطلت عنه خاصيته فى جذب الحديد فان اراد مريد ان يعيدها اليه انقعه ثلثة ايام فى دم تيس طري يجدد له الدم فى كل يوم فانه يعود الى خاصيته وقال غيره ان هذا الحجر الذي يجذب الحديد اذا دلك بالثوم انقطع جذبه فان القى فى الخل عاد الى حالته وقال ابو الفتح القاضي احمد بن مطرف فى كتابه فى الاحجار السبعة التي تماها الترتيب وقد ذكرنا المغناطيس وهكذا وجدته معتدلا بخطه فى كتابه المذكور فقال وانا جعلت هذا حجرا من هذا النوع فى درج طيب مع شيء من المسك والعنبر والكافور والند فبطل فعل الحجر بعد ان كان يجذب الحديد جذبا عجيبا فغاب عنه كاف ا وغسلته بالخل فما عاد يجذب شيئا وهو عندي الى الآن لا يفعل شيئا ٭

58

عن اعادته هاهنا معدنه الذي تكون فيه
العقيق يوتى به من اليمن فمن معادن له يعدن
وصنعا ويوتى به الى مكة شرفها الله تعالى ويوتى
بالعقيق ايضا من سواحل بحر رومه ومن مكة يجلب
الى ساير البلاد جيده وردية العقيق خمسة انواع
احمر ورطبي وهو احمر للصفره وازرق واسود وابيض
واجوده الاحمر ثم الرطبي ثم الذي يليه على الترتيب
الى اخرها خواصه ومنافعه العقيق حار يابس
وفيه ثلث خواص الاولى من تقلد بالاحمر منه سكنت
منه روعته عند الخصام والثانيه انه من تختم
بالنوع الثاني منه وهو الذي لونه لون ماء اللحم
اذا القي فيه الملح وفيه خطوط بيض قطع عن حامله
نزول الدم من اي موضع كان من الجسد لاسيما النساء

اللواتي

اللواتي يبدو من الطمث ۞ الثالثه انه اذا استيك باي
انواعه اتفق اذهب عن الاسنان صداها وبيضها ومنع
الحفر من الاسنان ان تخرج من اصولها الدم ومحروقه
يمسك الاسنان المتحركه ويثبتها اتصبه ۞ قيمته
وثمنه العقيق يصنع منه خواتم يباع الخاتم منها
باربعه دراهم ويصنع منه نصاب سكاكين يباع النصاب
بدينار فما دونه وقيمة الفص الجيد المنقوش درهم ۞
وهذا السعر كله انما هو واقع على الاحمر وهو الاول
من انواعه والرطبي دونه في الثمن واما بقيه انواعه
فلا قيمة لها يعبرها

الباب الثالث عشر

في الجزع ۞ علة تكونه في معدنه علة تكون
الجزع قد ذكرناها في علة تكون العقيق فاغنى ذلك عن
اعادتها هاهنا معدنه الذي يتكون فيه ۞

59

الجزع يوجد في معادن العقيق باليمن ومنه ما يوتى
به من الصين جيده ورديه ۞ الجزع انواع كثيره
منه البقراوى والعروى والفارسي والحبشي والعسلي
والعرق واما البقراوى فمنه حجر مركب من ثلث
طبقات طبقة حمرا لا تستشف لها يليها طبقة بيضا
لا تستشف ويلي الطبقه البيضا طبقه بلوريه تستشف
واجوده ما استوت عروقه في الخشن والرقه وكان
سليما من الخشونه وقبح العرض ووجود الاتانيه
واما الحبشي فانه عرقي وجهتاه العليا والسفلى
سود او كانت كالسبج والوسطى شديد البياض واجوده
ما كان من استواء العروق على ما وصفناه واما باقي
انواعه فاجودها ما اشتدت صقالته واستوت
عروقه بالحجر حجر ليس في الاحجار اصلب منه جسما

الا انه ذكر حسن سريعا وانما يحسن اذا طبخ بالزيت واذا
جلى على اليد بالعسل اشرق وانار ۞ خواصه في
ذاته ۞ ذكرت حكماء الفلاسفه ان الجزع انما
اشتق اسمه من الجزع لانه يولد الجزع في القلب وكذلك
قالوا من تقلد منه او تختم به كثرت همومه وراى في منامه
احلاما رديه مفزعه وكثر وقوع الكلام بينه وبين
الناس ولذلك صار اهل اليمن وملوكهم من حمير لا يرون
بشيء منه ولا يدخله خزاينها ولا تستعمل عندهم شامه
او يتقلد به الا اهل الجهل وعدم المعرفه بهذه الخاصيه
فيه واهل الصين يكرهون ان يحفروا معادنه وانما
يخرجه من بلادهم الزط الى بلاد الصين فيبيعونه وان
غلق منه على طفل كثر تسيلان لعابه خواصه
ومنافعه منها ما ذكره لي بليوس الانطاكي في كتابه

56

وباع الباقي جميعه على هذا السوم ورأيت بسوق القاهرة
المعزية كلأها الله حجارة كثيرة مدلسة مصنوعة
تباع على انها بازهر حيواني بسوم دينار المثقال وان
جربت على حلوق الافاعي لم تقتلها او على حمة العقرب
او غير ذلك من الامتحانات التي ذكرناها فيما سلف
لم يؤثر وربما وقع منه الخالص فيبيع بسعر المغشوش
لعدم الخبرة به وبذلك الامتحان له فصار سعره المتعارف
لاجل ذلك ما ذكرناه

الباب الحادي عشر في الفيروزج

علة تكونه في معدنه الفيروزج حجر نحاسي يتكون
في الابخرة النحاسية الصاعدة من معدنه على ما نذكره
من بعد في تكون غيره من الاحجار النحاسية معدنه
الذي [illegible] الفيروزج يجلب من معدن
له

له في جبل من جبال نيسابور ومنه الى سائر البلاد ومنه
نوع يوجد في نسا ولا ان النيسابوري خير منه
جيده ورديه الفيروزج نوعان سجاني وخيجي
والخالص منه العتيق هو السجاني والجوده الازرق
الصافي اللون المشرق الصفا الشديد الصقالة المستوي
الصبغ واكثر ما يكون فصوصا وذكر الكندي انه رأى
حجرا زنته اوقية ونصف خواصه في ذاته
منها انه حجر يصفو لونه بصفا الجو ويتكدر بتكدره
وذكر ارسطو ان كل حجر يستحيل عن لونه فهو
ردي للابسه ومنها انه اذا اصابه شي من الدهن
افسد حسنه وغير لونه وكذلك العرق يفسد ويطفي
لونه بالكلية وقد وقفت على ذلك منه بالتجربة وكذلك
المسك اذا افسده وابطل لونه واذهب حسنه

57

خواصه ومنافعه منها انه يجلو البصر بالنظر اليه
ومنها انه ينفع العيون اذا سحق في الاكحال ويشرب
ايضا الفيروزج للعارضة في الجوف ويجمع حجب العين
المنحرفة ويقبض نتو الحدقة والبثرة وينفع من غشاوة
البصر ومنها اذا سحق وشرب منه نفع من لدغ
العقارب وطبعه البرد واليبس ومنها ما نقل من
رسالة ارسطو الى الاسكندر في تدبير الملك وهو
اخر كلامه في الرسالة المذكورة ونصه خزنا حجر
حجر الفيروزج هذا لم تزل ملوك الاعاجم تتقلد به
وتستكثر منه وخاصيته العظمى انه يدفع القتل عن
ممسكه ولم ير قط في خاتم قتيل وهو اذا سحق وشرب
نفع من لدغ العقارب والهوام المؤذية المسمومة
قيمته وثمنه اكثر ما يوجد الفيروزج فصوصا

كما ذكرناه وفصوصه تختلف في الجودة والرداءة
اختلافا كثيرا فربما كان ثمن الفص درهما وربما كان
كان دينارا او درهما واحدا او متقاربه والاصل فيه ما
ذكرناه عند ذكر جيده ورديه والسجاني وامراء
بني ابن المغرب وقوادهم يطلبونه ويتغالون في ثمنه
وربما بلغوا بالفص منه عشرة دنانير مغربية يجعلونه
في حلية اسلحتهم ويتختمون به كثيرا والعامة يزعمون
انهم يدخلونه في صناعة الكيميا حتى ان الجزار زعم
ذلك وذكره في كتابه في الاحجار وليس لذلك بصحيح
وانما يتغالون فيه لاجل ما ذكر من خاصيته في دفع القتل

الباب الثاني عشر في العقيق

علة تكونه في معدنه قد ذكرنا في الباب الخامس
علة تكون العقيق وما قالت الحكماء فيه اغنى عن ذلك

54

والملك يحلف عليه ما يذهب منه ولا يفعل هذا الفعل
الا قوم مخصوصون بذلك واخبرني هذا المخبر
وغير من ثقات الفرس وعلمايهم وتجارهم واعيانهم
اخبارا متواترا متفقا عليه عندهم ان سلطان العجم
خوارزم شاه المذكور رام دخول ما يتاخم الصين
من بلاد الترك فلما قاربها توالت عليه امطار وبرد
ماء وثلج كان عسكره يهلك فيه ولم يكن اوان افراط
البرد بل كان قد تحرى حركته اليهم فصل القيظ لشدة
برد بلادهم فعلم انه فعلهم ما ذكرناه من الحجارة المذكورة
فبعث جندا ارفقه يطوفون الجبال فاتوه برجلين
وجدوهما يعملان ذلك بهذه الحجارة فلفهما في لبدين
اسودين ودفنهما جنبين فقشع جميع المطر والثلج والبرد
الشديد من حينه قال ورسم من يعمل ذلك اذا اعاد عليه

ان يسر

ان يفعل به ما ذكرناه والا لم يذهب اثر ما عمله مدة
طويلة قال واذا احمل في موضعين كان البرد والثلج
اشد حتى لا يطاق احتماله فسبحان من اودع
اسرار حكمته في مخلوقاته التي لا يعقلها الا العالمون
قال وهو اعز من قائل وكم من اية في السموات والارض
يمرون عليها وهم عنها معرضون واخبرني تاجر من
تجار الفرس المترددين الى بلاد الترك انه دخل مدينة
من مدن الترك تسمى طحان قال فالفاها اكثر اهل
الارض حيات قال وسقط عندهم من السحاب ثعابين
عظام وذكر انه اشترى بهذه المدينة قحف ثعبان
بخمسة وخمسين دينارا صنع منه نصب ثلث سكاكين
انواع الواحد منها بمائة دينار فباعه مشتريه الى ملك
من ملوك الفرس بخمسمائة دينار قال اصلحه الله اذا

55

قرب من طعام او شراب مسموم يرشح كله عرقا قال
وان ابار ودروره في هذه العظام ظاهرة قال
واذا اخرج منه عظم فراه كلب كلب او غير كلب
بهت ساعة يشخص اليه ثم يخر مغشيا عليه الى يومين
او ثلثة قال جربته بنفسي في ذلك قيمة البازهر
وثمنه اما الموجود المعدني بايدي الناس الان
فلا قيمة له يعتد بها لعدم الخواص والمنافع الموجودة
في البازهر الحيواني منه واما البازهر الحيواني فان
المعمول المدلس منه اكثر من الخالص بايدي الناس
وقد حضرت في دكان جوهري خبير بالاحجار من اهل
الاندلس بثغر الاسكندرية ودخل السوق رجل تاجر
اعجمي فاخرج ثمانية عشر حجرا على انها بازهر حيواني
ودفعها الى الدلال فاوقف عليها امين السوق فلم ينكر

منها شيئا ونادى عليه جملة على انه بازهر حيواني
فلما وصلت الينا وراها الجوهري الذي كنت في دكانه
اخرج منها حجرين فاراني ايهما واخبرني ان ليس في الجميع
بازهر غيرهما خالص وان الباقي معمول مدلس واستدل
على صحة قوله بامارات ابرزها في المعمول وغير المعمول
تظهر للذكي اللطيف لنظر الجيد الفطنة فان المصنوع
لا يكاد يخفى عن الفطن اللبيب والذكي الاريب فطلبنا
من العجمي من الجوهر بيع الحجرين دون ساير الاحجار فامتنع
وقال لا ابيع الا الكل جملة كما اشتريتها فلما جهدناه
به ذلك ولم يفعل خاونا به واعلمناه ان الحجارة مدلسة
مدلسة مصنوعة سوى الحجرين واوقفناه على صحة ذلك
فانكسر عند ذلك وقال هكذا اشتريتها وحده واحجرا
واحدا ابيتم فاخذت احد الحجرين بخمسين دينارا المثقال

52

الحجارة التي يجلب بها المطر اخبرني من اثق به من
الثقة الواقعة في ذلك انه شاهد ببلاد الفرس بين
بخارى وسمرقند في عسكر الملك المرحوم سلطان العجم
علاء الدين محمد خوارزم شاه رجلا تركيا يعمل بالحجارة
الذي نذكرها فينزل المطر الغزير في اي وقت شاء
السلطان قال وذلك ان بلاد الصين والفرس
طايرا يسمى جراف وتفسير هذا اللفظ اجمل الما وذلك
ان سرح في لغتهم اجمل وان الما قال وهذا الطاير
كالاوزة الكبيرة احمر الريش قال وهذا الطاير الذي
يسمى بدبانه البشمور بعينه وهو مضر كثيرا
يعلقون ريشه للزينة في المراكب قال وهذا الطاير
ببلاد الصين وما يتاخمها وكثير ببلاد الفرس بعش
في جزيرة مياة منقطعة في زمن الامطار والسيول فاذا

نضب الماء ينبعث من اعشاشه فيحفر تحت عشه قدر
ذراعين فيوجد هنالك حجر واحد في قدر البيضة الكبيرة
لونه غبري فيه نكت بيض وحمر نحو الجمشت فكل ما كان
ارخى كان اجود فيجمعها قدر على من هذه الحجارة
ويرفع الى خزانة الملك وتوضع في صندوق تحت يد امين
مكلف بذلك مفاتيحه تحت يده فاذا اشار الملك
في الصيف في بعض اسفاره واذاه الحر وغبار الطرق
في فصل القيظ او اتفق غير ذلك من الاسباب التي
يحتاج فيها الى كثرة الما ورطوبة الهواء امر الامين
المكلف بهذه الاحجار باستعمالها فاخبرني من حضر
كيفية العمل بهذه الاحجار من اهل غزنة في معسكر
السلطان المذكور قال شاهدت شيخا من الترك حضر
واقيمت له خركاه يستتر بها عن الناس قال وحضرت

53

انا والامين على الحجارة المذكورة قال وكان على الخركاه
مفتوحا الى جهة السما غير مستور ثم وضعت بين يديه
طاسة فيها ما ثم اخذ ثلاث قصبات غلاظ فاقام
احدها الى جانب الطاسة الايمن والاخرى الى
جانبها الايسر ثم مد الثالثة معترضة على الطاسة
طرفاها على القصبتين القايمتين ثم اخرج ثعبانا
رقيقا لونه كلون الحجارة اغبر منقط بحمرة وبياض
فربط ذنبه بخيط وعلقه في القصبة المعترضة
منكسا راسه فوق الما بقدر ذراعين ثم اخذ حجرين
من الاثنين من الحجارة المذكورة فوضعها في الما
ثم رفعهما فحك احدهما بالاخر حكا يسيرا ثم رماهما
في الما ثم حكهما ثم رماهما سبع مرات فعل ذلك
ثم اخذ من ذلك الما فرش منه على الارض قال

وكان الذي يعمل هذا العمل مكشوف الراس وكذلك
من شرطه ان لا يستر راسه وقت عمله هذا العمل بل
يكون مكشوف الراس محلول الشعر مقطب الوجه
كالغضبان وهو في خلال عمله يومي براسه الى
السماء ويتكلم بكلام يستدعي به المطر يفعل ذلك
مقدار ساعتين من الزمان فلا يلبث ان تغيم السما
ويأتي المطر الغزير العظيم قال هذا المخبر رحمه الله
هذا الفعل لا شاهد عجايب صنع الله واسراره المودعة
في مخلوقاته فاجي والسما صاحية فلا انصرف الى محلي
الا في الطين والسيول وذلك في حرارة القيظ قال
واخبرني هذا الشيخ التركي المتولي لهذا العمل انه كل
مرة يفعل هذا يصاب في اهله وينوبه اما موت ولد
او ذهاب مال او ما اشبه ذلك وانه لا يبرح محتاجا

50

فيه بردة واكلناها وبمنا في ظل شجرة نشرنا عليها اللحم
المقدد ثم استيقظنا ونحن جياع ووجدنا اللحم قد جف
جفافا لم نعهده يجف مثل ذلك الجفاف في مثل ذلك الوقت
الذي بمنا فيه فحملناه واتينا به المدينة وكان اليوم الذي
بمنا فيه ضحى يوم الثلاثا فوجدنا الناس خارجين من صلاة
الجمعة ولم يكن بين الموضع الذي بمنا فيه وبين المدينة
الا مسيرة يوم فصح عندنا انا بمنا نحو ثلاثة ايام بلياليها
في نومة واحدة واخبرني هذا الصياد ان العقعق
المذكور يقتل بالتخدير الشبيه بالنوم بتأثير اكلها دون
ان يحس في جسده بشي من الالم البتة نوما متصلا لا يقوم
منه وذلك انه يحل في بدنه دم قلبه فيموت وهو لا
يحس بالالم البتة قال مصنف الكتاب وقد
اكلت اللحم الابل سفينة عند مامرة الهمنية هذا الصياد

المخبر

المخبر لي بهذا الخبر وغيره وكان يهدي الي منفرد اراكط
منه وهو لحم لذيذ طيب وفي الاحجار التي توجد في
بطون الحيوانات خواص عجيبة من ذلك ما ذكره ياسقور
يدوس من ان الخطاف اذا اخذ فرخه في زيادة القمر
وكان اول ما افرخ وشق جوفه وجد فيه حجارتان
احديهما ذات لون واحد والاخرى مختلفة اللون فان
شدا في جلد عجل او جلد ايل قبل ان يصيبهما تراب
وربطا على عضد من يصرع او رقبته انتفع بذلك ثم قال
وكثيرا ما فعل مرارا فابرا من به صرع برءا تاما وذكر
هذا الحجر للاسكندر تعالى انه يوجد في جوف فراخ الخطاطيف
في اعشاشها حجران ابيضان او ابيض واحمر فالاحمر
اذا علق على من به صرع ابراه والابيض اذا وضع على
مصروع اقامه وان علق عليه لم يصرع وكان لين القطن

51

في حرز اليرقان انها صفرا معروفة تعلق على صاحب
اليرقان فينتفع بها نفعا عظيما وهي توجد ايضا في اعشاش
فراخ الخطاطيف وقال هو وغيره من العلما بالخواص
ان الخطاطيف كثيرا ما يعتري فراخها اليرقان وانه
اذا اصابها ذلك ذهبت كبارها فاتت بتلك الخرزة
فتضعها عندها فيذهب عنها اليرقان وان هذه الخرزة
اذا طلبت في اعشاشها ولم توجد احتيل لحصولها بان
يعمد الى فراخ الخطاطيف في غيبة كبارها فتطلى بزعفران
مذاب بالما فاذا اتت الخطاطيف ورات ذلك ظنت
ان اليرقان عرض لفراخها فتذهب وتاتي بالحجر وتلقيه
في اعشاشها فتوخذ منها اذا ذهبت وذكر ارسطو
في كتابه في الاحجار انه يوجد في باطن الديوك حجارة
منها الى البياض ومنها الى لون المها قال فاذا اصيب

منها شي فاخذته وعلقته على المجنون برا وان علقها الشاب
على فخذه زاد في الباه وكثر في الجماع ويطرد عنه
الشيطان وكل ريح السوء وفزع الصبيان الذين
يفزعون في النوم ونصر باسنانهم وفي مرارة البقر
خاصية نافعة وذلك انه يوجد فيها شي كالحجارة يوخذ
منه ويسعط به من به ضعف البصر ورقته ومن يخوف
عليه اجتماع الما في عينيه ويسعط منه ايضا لمن
يصيبه الصرع الشديد ويسعط منه بقدر العدسة
مع ما السلق المعصور من اصوله فينتفع به نفعا جيدا
وقال احمد بن ابي خالد المعروف بابن الجزار في كتابه
في الاحجار ومن سحق الحجر الذي يتولد في المثانة في الكلى
والمثانة وخلطه مع الاكحال نفع بياض العين نفعا بينا
ومن المشهور عند اهل المشرق والمتفق على صحة نقله

48

من الباهر وهو الموجود في قلوب الايايل افضل من جميع
الاوصاف المذكوره في الباهر حتى انه اذا حك بالماء
على مسن وسقي منه كل يوم وزن نصف دانق للصحيح
على طريق الاستعداد والتقدم بالحوطه قاوم السم
القتال وحفظ من مضاره ولم يخش منه غايله ولا اماره
خلط كما يخشى من المثرود بطوش ولابة المحرورين
ولا المحرسين لانه انما يفعل ذلك بخاصيه جوهره
قال وفي طبع هذا الحجر حراره يسيره وخاصيته
النفع من ساير السموم الحيوانيه والنباتيه الحاره والبارده
ومن خواصه ان من تختم منه بوزن شعيره في فص خاتم
ثم وضع ذلك الفص على موضع اللدغ من العقارب والهوام
الطيارات وغير الطيارات ذوات السموم و[illegible]
الزنابير والدراريح نفع نفعا بينا ومن خواصه انه ان
سحق

سحق ونثر على موضع لدغ الهوام الارضيه حين تلدغ جذب
السم واستخرجه وان نغض الموضع قبل ان يبادر البلل وانا
ثم نثر عليه من هذا الحجر مسحوقا ابراه ومن خواصه
ما ذكره بعض الحكماء من الاوايل انه اذا اصنع خاتم ذهب
يكون فصه باذهر ونقش عليه صورة عقرب حين يكون القمر
العقرب سعد العقرب وبد من اوتاد الطالع وعملت
اقراص من كندر وطبعت بالخاتم ورفعت لمن لسعته
العقرب وشرب قرصا من هذه الاقراص المختومه بهذا الفص
الباهر لم تضره اللسعه وبرا منها قال احمد بن يوسف
كاتب ابن طولون مفسر كتاب الثمره لبطليموس وقد
جربت هذا الفص وختمت بهذا الخاتم على غير الكندر فعل
كما يفعل اذا ختم الكندر ومن خواصه انه اذا كان
منه فص خاتم مع انسان ثم لدغ هو او غيره او سقي سما

49

ووضع المسموم ذلك الفص في فمه ومصه انتفع به وان لم يفقص
من الفص قال المصنف للكتاب ومن عجايب اثار
الله وصنعه في الحيوانات ان الايايل التي في المواضع المذكوره
في ما يلي من بلاد الصين يستخرج منها حجر الباهر النافع
للسم كما ذكرناه والايايل الموجوده في جهات جميع المشرق
والمغرب كلها تستخرج منها السم القاتل الوحي وذلك
ان العقد التي في ذنب الايل الموجود في ذنب المشرق سم
قاتل واظن ذلك ايضا موجودا في ايايل الباهر فهذا السم
يقتل بالتخدير وجمد الدم وقد شاهدت من ملوك المشرق
والمغرب من كان يرضي الصيادين عليها اذا دفع لهم
ايل يأخذونه منه معونه على خزانة الملك ومن
اعجب عجايب صنع الله في اسرار حكمته ان هذا الحيوان
واعتذاؤه بالسم القاتل لكل حيوان على الاطلاق

وسريانه في جسده دون اضرار به ثم اجتماع بيد من
السم نفسه في طرف ذنبه وضد السم وما ينفيه في
موضع اخر من جسده ومع ذلك فلحمه يغذو غذاء صالحا
لا مضرة فيه البته وقرنه اذا تبخر به لم يعرف الموضع
الذي يتبخر به فيه حيوان ذو سم مجرب فتبارك العزيز
العليم القادر على ما يشاء الفعال لما يريده واخبرني
صياد ممن كان يصيد الايايل بالمغرب ويستخرج
عقد اذنابها لبعض ملوكها حكايه عجيبه في الايايل
وذلك انه قال وقع لي ايل سمين وذلك بافريقيه بقرب
من مدينه قفصه وكان معه نفر من اصحابه فاستخرجوا
تلك العقده وزوجوها وقددوا لحم الايل ليحملوه الى
الابنيه كما جرت العاده قال ثم طبخنا من اطايبه
ومن سمينه قدرا كبيره فكان عليها دهن كثير وتردنا

46

كبراء المغرب ممن حضر مجلس مليكه الاعظم يعقوب
بن يوسف بن عبد المومن انه احضر ذات يوم سفطا مملوا
حجارة واشار على رؤساء الاطباء وامناء الجوهريين وقال
لهم ان هذه حجارة بازهر اجتمعت عندنا فامتحنوها
فاذا صحت فرقوها على امناء الاسواق ومشايخ الارباع
بجميع المدينة ليشتهر امرها وتكون عندهم للثواب
تبذل للمحتاجين اليها يستعمل منها قدر الحاجة للملدوغ
وقت الحاجة الى ذلك قال فعددت فكانت ما بين
وثمانين حجرا فحضرت الاطباء وصيدت الافاعي وارسلت
الى الفراريج بلدغها وحلت الحجارة اظهرت الفراريج
فلما خلص من الموت صح حجره ولم يخلص بطل حجره فصح
من الجملة ستون حجرا وكان الباقي مدلسا ونرجع بعضا
اليه وفرق الباقي كما ذكر خواص البازهر الحيواني

في ذاته منها ان قرب للعقرب بطل لسعها وان لسعت
لم يؤذ سمها ومنها ان افواه الافاعي وحلوقها اوسع من
اجوافها في اصل خلقتها قدر شعيرتين واذيب بماء
وصب في افواه الافاعي والحيات خنقها وماتت وهذا
الذي قبله مما يمتحن به البازهر الخالص من المغشوش
ومنها انه اذا جعل في اجسام خشنة مباشرة لجسمه
بحكة معه تغيرت صورته واكلت منه خشونته
واذهبت نضارته وبدلت جميع صفاته حتى لا يكاد
يعرف وقد كان عندي حجر بازهر حيواني خالص فجعلته
في كيس فيه ذهب دنانير ثم سافرت سفرا بعيدا فلما
استقريت فتحت الكيس واخرجت حجر البازهر فلم اعرفه
حتى ظننت انه قد عدل على التغير جميع صفاته ثم وزنته
وجدته اقل مما كان فيه فزاد تشككي ولم يكن معي

47

من اتهمه فعجبت من ذلك وبقيت متحيرا في امره ثم جعلته
في حق صغير بعد ان لففته بابريسم وغفلت عنه مدة
ثم اخرجته وجدت الحجر الذي كنت اعرفه اولا وقد زالت
عنه الهيئة الرديئة التي اكتسبها من احتكاكه بخشونة
الذهب ورجعت اليه صفاته الاولى الا ان وزنه نقص
بما احتك منه في الكيس بخشونة الذهب ولما كان بعد
ذلك جرى ذكر البازهر بيني وبين حذاق الجوهريين فعرفني
ان من خاصيته التغير بالاحتكاك بالاشياء الخشنة
فحينئذ بما شاهدته في ذلك بالتجربة تصديقا لقوله
خواص البازهر الحيواني ومنافعه البازهر اسم
اعجمي فارسي اصله في لغة الفرس باد زهر فباد
بالفارسية النظافة وزهر السم ومعناه منظف السم من
الجسد فلما عرب اسقطت الدال فقيل بازهر

واخص خواص هذا الحجر النفع من السم اي سم كان من
السموم القاتلة من حيوان او نبات او جماد من السموم
الحارة والباردة ومن عض الهوام ذوات السموم ومن
لسعها ولدغها اذا شرب منه اربع شعيرات الى اثنى عشر
شعيرة مسحوقة او محكوكة بالمبرد او محكوكا على المسن
بزيت الزيتون او بالماء فانه يخرج السم بالعرق من بدن
المسموم ويخلص نفسه من الموت ويفعل ذلك بجملة
جوهره والخاصية الموجودة فيه وهو حجر شريف نفيس
ليس في جميع الاحجار ما يقوم مقامه في دفع السموم
ومن خواصه انه اذا سحق وذر على مواضع النهش
واللدغ من ذوات السموم جذب السم الى خارج وابطل
ضرره ومن خواصه ما ذكره ابن جميع في كتابه
المعروف بالارشاد قال ما نصه هذا قال هذه والحيواني

44

لنا ما قدمناه في اكله للحيات وخروج البازهر من
عروقها لاجل ذلك قال فاجتهدت في صيد ايل فوجدت
في اماقيها رطوبة مجتمعة بعضها بقدر الحمصة وبعضها
بقدر الفولة فاخرجتها وجمعتها فاخرج لي ما جمعه من
ذلك واريته فرايته حجرا الى السواد غير منحجر رطبا
كالشمع فسالته هل جربته في شيء من السموم فقال لا لكن
صح عندي ان هذه الايايل تاكل افاخ الحيات ممن شاهد
ذلك عيانا من الصيادين وغيرهم وكشف لي هذا الامير
المذكور عن عضده فوجدت على عضده حاشية حرير
مربوط فيها صرة صغيرة مستديرة في قدر البندقة الكبيرة
او اصغر فيها حجر وذكر ان الملك الاشرف رحمه الله
زوجه ووهبه اياه وذكر انه ابيض حجر من نوعه في الحجوم
الاحجار ان احد مما باقي عند الملك الاشرف والاخر عند

الخليفة ببغداد فسالته عن خاصيته فقال
ان من نهشه حيوان مسموم فوضعه على موضع النهشة لصق
بالموضع واجتذب السم واخرجه بالرشح فلا يزال عالقا الى ان
لا يبقى من السم شيء البتة فيبرا الملسوع ويسقط حينئذ وذكر
انه جربه في ذلك فيمن سقي سما فوضعه في فيه فجذب
السم واخرجه بالقيء وذكر انه لا يعرف له اسما ولا اصلا هو
معدني او حيواني القول الثاني ان هذا الحجر يتكون في قلب
هذا الحيوان وانه يصاد لاجله ويذبح ويستخرج الحجر
من قلبه وهذا القول لم اره الا لمتاخري اطباء مصر
وهو غير صحيح والقول الثالث ان هذا الحجر يتكون في
مرارة هذا الحيوان كما يتكون كثير الاحجار في مرارات
كثير من الحيوانات بالمشاهدة فيصاد هذا الحيوان ويذبح
ويخرج البازهر من مرارته وهذا عندي هو الصحيح

45

من امر البازهر الحيواني المجلوب من بلاد الصين
وقد اخبرني بعض العلماء الثقات من مشايخ الجوهريين
المعول على قولهم الموثوق بصحة نقلهم ان الخالص من هذا
الحجر الحديث المجلوب ذا الجنس ظاهره باللسان وجد طعم
المرارة فيه ظاهر الذوق واخبرني هذا المخبر انه كسر
حجرا منه فوجد فيه حشيشة اشتمل عليها الحجر من اصل
تكونه في معدنه الذي يتكون فيه معدنه الذي يتكون
فيه اما الحيواني فقد علمت من كلامنا فيما سلف حيث
يكون من الحيوان وحيث يكون حيوانه من البلاد ه
وكذلك المعدني ذكرنا حيث يكون صنفاه ه
[illegible] الخالص الجيد النافع من البازهر الموجود
في عصرنا هذا انما هو الحيواني لا المعدني وهو اصفر قشري
خفيف واغبر منقط نقطا صغارا خفيفة وابيض اللون

من المذاق واجود المعدني هو الصيني وهو حجارة صغار
صفر اترجية فيها طرائق خضر وقد ذكر ارسطو البازهر
المعدني وقال ان اصنافه كثيرة ومنه الاصفر والاغبر
والمشوب بشيء من الحمرة والمشوب بياضا وذكر لي ان
اجوده الاصفر وان البازهر الحيواني كثيرا ما يغش ويدلس
فيصنع حجارة صغارا من اجساد مجموعة تشبه اذا جفت
شكل البازهر الحيواني وكثيرا ما يغش والاختبار الذي
يفصله عن البازهر الحيواني انه اغبر كمد اللون شاذ
غير منقط اللون والخالص اصفر واغبر بصفرة منقط
نقطا صغارا كالنمش وطبقاته ادق وكبارا من طبقات
المصنوع المدلس وهو اخف وزنا واهش مكسرا ومحكه
ابيض كاللبن الحليب او قريبا منه والخبرة الصحيحة
فيه الامتحان بالتجربة في السم واخبرني بعض

42

الباب العاشر في البازهر علة
تكونه في معدنه قال احمد الموجود من هذا
الحجر بايدي الناس صنفان احدهما حيواني والاخر
معدني واما المعدني منه فاني وقفت عليه بمعدنه
بنفسي بالتخوم بين بلد جزيره بن عمر وبلد الموصل
وهو هناك كثير ويوجد منه حجار كبار يتخذ نصبا
للسكاكين وغير ذلك تبلغ القطعه من اوقيتين واكثر
من ذلك وهذا النوع منه ابيض وفيه نقط من الوان
اخر صفر وغير ذلك من الالوان وليس لشي منه
نفع في السموم اصلا وهو حجر رخو المجس اسم الحكاكه
سريع الانحكاك ليس له خاصيه ولا نفع الا انه يحل
بالما ويطلى بمد انه موضع الضربه او السقطه الوارمه
المتغيره اللون الى زرقه وسواد فيبريه ويفش ورمه

ويزيل الدم ويرد الدم الى اللون الطبيعي في اسرع وقت
واجاه حتى انه يكاد يتاخر فعله في ذلك عن يوم او
ليله لا اكثر جربت ذلك بنفسي غير ما مره فيه فوجدته
كما ذكرته لك واخبرني المعدني الذي ذكرته في باب
الزمرذ انه في معدن الزمرذ الذي بمصر شي كثير
جدا كبير الجرم لدنه نصاب للسكاكين واكثر من ذلك ومن الموجود
الان من البازهر المعدني نوع يجلب من الصعيد حجار
صغار اصفر سادج وبرش منقط نقطا صغارا الوانه
مختلفه ينفع حكاكه لذغة العقرب لا غير منفعه
صغيره فاما البازهر الحيواني فهو المقصود في هذا الكلام
في هذا الباب وهو حجر خفيف هش اصفر واغبر منقط
نقط خفيفه يوجد طبقات رقاقا في اصل تكوينه
طبقه فوق طبقه يوجد الا كذلك ويحل سريعا

43

اذا حك ويحكه الى البياض واعظم ما يوجد منه مثقال
الى سبع مثاقيل يوتى به من بلاد فارس من تخوم الصين
والحيوان الذي يوجد فيه هو الايل الذي يكون بتلك
البلاد والحيوان الذي يوجد فيه البازهر يشتهي اكل
الحيات ذوات السموم القتاله لا سيما ما صغر من اولادها
وهي مطعم غذايه يبحث عنها ويستخرجها من حجرات
فياكلها وقد اختلف الناس في اي موضع من جسد
هذا الحيوان تكون البازهر على ثلاثة اقوال القول
الاول تكون في عينيه بالدا وذلك انه اذا اكثر اكله
اولاد الحيات اعتراه حكه في جميع جسده من سمها
فيعمد الى بركة ما فيغوص فيها رافعا راسه على الما الى
ان يغيب كله في الما حتى لا يظهر منه الا انفه فتنبع
حينيذ من جسده بخار حار رطب الى عينيه ثم يخرج من مايه

الذين يليان انفه يمنه ويسره ويستحيل ما فاذا اضر بهما الهواء
جمد وتجسد حجرا معلقا بشعر لحيتيه حتى يعرض
مثل ذلك العارض فيفعل مثل ذلك الفعل فيخرج حجر
اخر ويستحيل ما ويسيل من ذلك الموضع فيعليه على
الحجر المتكون قبل فيجمد اذا باشره الهوا فوق الحجر الاول
كما جمد الذي قبله فلا يزال كذلك دايما حتى يثقل الحجر
فيسقط من ذاته ويحكه الحيوان اذا ثقل عليه الى حجر
او اصل شجر فيسقط في مكانه حتى يوجد فيوخذ واخبرني
الامير الكبير الفاضل العالم سنقر بن قليج اعلا الله
جل وحرس مجده انه شاهد بعينيه في بعض سفراته
في التخوم بين بلد حلب وبلد الروم لموضع يسمى مرعش
وبلغ ما به ايايل تغوص في بركة هناك لا يظهر
من الايايل الاجبله فساله نسال عن السبب في ذلك فقيل

40

فهو كناظر الهر حقيقه ولذلك سمي الجزع عين الهر وان
كسر الحجر او قطع على الاجزاء ظهرت تلك النكته في كل
جزء من اجزايه واجوده ما اشتد بياض ابيضه
وشفيفه وكثرت مايته النكته التي فيه وحقت حركتها
وظهر نورها واشراقها وكان اذا اشرق وهو ساكن
رئي فيه ماء كالموج متحركا كاشف مايكون حتى
يلقى بونه على مايليه فان كمل ترادت حركته تموجه
حتى يظن ان فيه ماء وليس فيه ماء وانما هو جوهر
شفاف يظهر فيه كله واحسنه المسقف الحسن
الشكل الكبير الجرم وتزيد في جودته والغبطة به
كتابه الاحجار خواصه ومنافعه اذ لم اجد له
ذكرا في كتاب من كتب الاحجار لا متقدم ولا متاخر
واظنه محدث الظهور ربما يلقى في الناس الا ان المشهور

الجزاوا

خجاب
المتبدا وان من خواصه عند جمع من اتينه من علماء الا
انه يحفظ حامله من الاعين السوء والانفس الخبيثة
الظاهره والخفيه والانسيه والجنيه ومما انقله فيه
من ثقات الجوهريين ممن دخل الهند وحوله فيه لطلب
عجب خواص الاحجار والوقوف على غرايبها ان هذا الحجر
يجمع خواص الياقوت البهرمان في منافعه باسرها
ويزيد عليه بخاصتين عظيمتين احدهما انه لا
ينقص ما لحامله ولا يعتريه فيه النكبات والافات
على اختلاف اسبابها والخاصيه الاخرى انه اذا كان
وخصم مصاف حرب ثم هزم حزبه وراى نفسه
لا ينجيه الفرار القى نفسه بين القتلى راه كل من هم
به من اعدايه انه مقتول متشحط في دمه فتنفر عنه
النفوس حتى لا يقربه بشره واخبرني بعض من دخل

41

الهند من الجوهريين انه راى هذا الحجر في المعابد تعبد
كما تعبد الاصنام قال وثمنه عندهم اغلا من ثمنه
ببلاد العرب وهم به اغبط وهو عندهم اعز وذكر
انه وقف على حجر منه بيع في المغرب بمايه وخمسين دينارا
لعله لا يساوي في عين الهند عشر هذا الثمن وذلك
لعلمهم من اسرار خواصه مما يجهله غيرهم من الناس
ووقوفهم عليها بالتجربه قيمته وثمنه هذا الحجر
يختلف فيه بحسب وقوع الشهوه فيه والعلم بخواصه
الا انه اذا وقع ببلاد العرب ولم يطلب بسوى المقال
منه خمسه دنانير وتزيد على ذلك بحسب وقوع الشهوة
فيه والعلم بخواصه كما ذكرناه وهو ببلاد الهند
وما يتاخمها من بلاد العجم اغلا ثمنا اضعافا مضاعفة
منه بغايرها من بلاد العرب لعلمهم بسير الاحجار

التي ترخص اثمانها بالقرب من معادنها وتغلي بالبعد منها
واخبرني رجل من اهل غزنة انه راى حجرا منه بيع
ببلخ بسبعمايه دينار وحدثني رجل من اهل عدن
لا اشك في صحة نقله انه توفي بها رجل يعرف باسمعيل
السلاوي في عام احد واربعين وستمايه فالقي في تركته
كيس فيه فص عين هر زنته مثقال او يقاربه ومعه
رقعه مكتوب فيها شرا هذا الفص من المعبر بسبعمايه
فيلي والفيلي عندهم مثل الدينار الذهب اليعقوبي
وثمنه باليمن خمسه ملكيه وهي دينار وربع مصري
قال فاشتراه صاحب اليمن من رسول من تركه المتوفى
المذكور بسبعه الاف ملكي وزعموا انهم طلبوا فيه وعشره
وذكروا ان هذا الفص كانت له مايته يلقى الشعاع
ويتموج كما ذكرناه فيما سلف ه

بلغ مقابلة

38

صباغ الذهب فان همز دوه وقعت تلك الحبة
في مباردهم وافسدت المبارد وملكها ومن خواصه
انه يثقب الدر والياقوت والزمرد وغيرها من جميع
ما يعمل فيه الحديد من الاحجار كما يثقب الحديد الخشب
وذلك ان تركب في راس خشبة حديدة وتجعل في
طرف الحديدة منه قطعة بقدر ما يراد من سعة
الثقب وضيقه ثم يثقب به فينثقب بسرعة والماس
بارد يابس في الدرجة الرابعة خواصه ومنافعه
منها ما ذكر ارسطو وجرب فصح من انه من كانت
به الحصاة الحادثة في المثانة ومن مجرى البول ثم
اخذ حبة من هذا الحجر والصقها في مرود نحاس او فضة
بمصطكي الصاقا محكما ثم ادخل ذلك المرود الى
الحصاة ولقيها فتفتت تلك الحصاة الخ

احمد بن ابي خالد المعروف بابن الجزار في كتابه في
الاحجار وبهذا الفعل عالجت انا وصيفا الخادم
صاحب المظلة من حصاة عظيمة كانت به وامتنع من
الفتح عليها بالحديد فلما فعلنا به هذا الفعل تحللت
الحصاة حتى صغرت وسهل عليها خروج ما بقي من
البول ومنها ما ذكر ارسطو في كتابه في الاحجار
ان الماس ينفع من المغص الشديد ومن فساد المعدة
اذا علق على البطن من خارج ومن كتاب ابن صهار بخت
انه حجر صلب جدا بارد يابس في الدرجة الرابعة وقيل
حار يابس ولذلك يجلو الاسنان وينقضها ويثقبها
وان علق على الطفل في وقت ولادته حفظه من الصرع
والفزع واذا ادخل في الفم اثبت الاسنان وربما
قتل اذا كان قد بقي عليه شي من سم الافاعي لانه انما يوجد

39

في وادي الافاعي ن قيمته وثمنه القيمة الوسطى
المتعارف في الماس زنة قيراط بدينارين وذكر
يعقوب بن اسحق الكندي في كتابه في الاحجار ان قدر
ما عاين من هذا الحجر ما بين الخردلة والحمصة وانه لم
ير منه اعظم من ذلك وذكر ان غاية ما شاهد منه
ببغداد المثقال بثمانين دينارا واول اذا انكسر منه
قطعة كبيرة تصلح للفص قدر نصف مثقال تضاعف
ثمنها على ثمن الحجر الذي يوجد على قدر الخردلة او قدر
الفلفلة ثلاثة اضعاف واربعة وخمسة وذكر ان ارخص
ما شاهد منه ببغداد المثقال بخمسة عشر دينارا

الباب التاسع في عين الهر

علة تكوينه في معدنه هذا الحجر ابتدا في معدنه ليكون
ياقوتا فبعدته عن الياقوتية الاعراض المغيرة له كما

اقعدت الماس وغيره من الاحجار التي ابتدات لتكون
ياقوتا بالزيادة والنقص من الكيفيات الاربع الا ان
الرطوبة والمائية اغلب منها على الماس وغيره من
الاحجار والياقوتية معدنه التي يتكون فيه
هذا الحجر يوجد في معدن الياقوت مع الماس فيما
سلف جبل سرنديب هذا الحجر عجيب وذلك ان
الغالب على لونه البياض باشراق عجيب ومائية رقيقة
شفافة الا انه يرى في باطنه نكتة الى الزرقة ما هي
على قدر ناظر الهر الناظر النور المتحرك في فص مقلته
وعلى ذلك اللون سوا وتلك النكتة مع ذلك
متحركة على الدوام اذا حرك الفص تحركت الى خلاف
حركته بحيث ان اميل جهة اليمين مالت النكتة متحركة
الى جهة الشمال بالعكس وكذلك في الاعلا والاسفل

36

الماس نوعان الدبشي والبلوري والزيتي اجودها
والبلوري ابيض شديد البياض كلون البلور والزيتي
يخالط بياضه صفرة كلون الزيت وهو شبيه بلون
الزجاج الفرعوني واخبرني بعض تجار الجوهريين
من العجم المترددين الى بلاد الهند والصين لاقتنائه
نفائس الاحجار ان في الماس نوعا له شعاع عظيم
اذا ظهر لقي شعاعه على ما يقرب منه من حائط
او ثوب او وجه انسان او غيره بنور مختلف الصور
اشبه شي بقوس قزح فان هذا الصنف من الماس
تحلى اكابر اهل الهند حليا يلبسونه للتجمل به ولا
يسمحون باخراجه من ايديهم البته وما لم يلق الشعاع
منه فهو الذي يستعملونه في قطع الياقوت وتجزيته
الى التجار [illegible] ومن خواص
الماس

الماس ان جميعه ذو زوايا قائمة ست زوايا او ثمان
او اقل من ذلك او اكثر محيط بزواياه سطوح قائمة
مثلثة الشكل واذا كسر لا ينكسر الا مثلثا ولو كسر على
اقل الاجزاء ومن خواصه انه يقطع كل حجر يمر عليه
وهو في نفسه مع ذلك عسر الانكسار حتى انه ان وضع على
سندان حديد ودق باعظم مطرقة لم ينكسر بل يدخل
في وجه السندان ووجه المطرقة ويخرقهما وانما ينكسر
بان يصير في شي من القمع ثم يدخل في انبوب قصب
وينقر بمطرقة رصاص برفق ومداراة يجب لا يباشر
جسمه الحديد او يضرب في انبوب اسرب وينقر برفق بشي
صلب فانه ينكسر وهو حجر صلب يأكل الحجارة كلها
كما ذكرناه حتى لا يلتزق بشي من الاحجار [illegible]
واذا الحجر [illegible] النقوش [illegible] في

37

الاحجار كلها ثم انه مع جميع ما وصفناه لا يكسره شي ويهشمه
الا ارخاها والينها واضعفها واكثرها زخاوة وضررا
الذي يوثر فيه تلك الاحجار ولا يوثر فيها وهو الاسرب
ومن ها هنا استدلت الحكما بان الماس حجر ذهبي
لانقياد الاسرب بحجر الماس كما ذكر دون الاحجار
كما يفسد الذهب ومن خواصه ان الذباب يشتهيه
فاذا ترك من قطعة صغيرة سقط عليها الذباب
يبتلعها ويطير بها الى [illegible] احمد وقد ذكر في خواصه
ان الانسان اذا ابتلع منه قطعة ولو كانت اصغر
ما تكون خرقت امعاه فقتلته على الفور واخبرني
بذلك ثقة من الجوهريين ايضا بالمغرب وذكر انه
شاهد ذلك ثم اخبرني جماعة من الجوهريين بالمغرب
والمشرق انهم كثيرا ما يجعلون القطعة الصغيرة

منه في افواههم فتشري مع حلوقهم ولا تضرهم شيئا بحثت
مع الصحيح في ذلك فثبت عندي ان النقلين صحيحان
معا وسببه ذلك ان الماس توجد في واد كثير الافاعي
فيسيل لعابها عليه فمنه ما يسيل لعاب الافاعي
جامد عليه فاذا ابتلع قتل على الفور وان قعد على
الفم ولم يبتلع اثر في [illegible] والحلد وما
وصل [illegible] عليه شي من لعاب الافاعي اما لانه لم يكن
عليه ذلك في معدنه بالاتفاق او لانه ذهب عنه
بسبب من الاسباب فانه لا يفعل شيئا من الافعال
المذكورة فيظن لذلك انه جرى منه وسبب ذلك
في صحيحه ما ذكرناه ومن خواصه ما يذكر ارسطو
من ان بينه وبين حجر الذهب محبة شديدة فيتشبث بالذهب
حيث كان حتى يخالطه منه الحته الخفيفه له وذلك

34

اشتهين الجماع شهوة شديدة لا يملكن انفسهن معها
خواصه ومنافعه من تختم منه بوزن عشرين حبة
من شعير لم ير في منامه احلاما رديه مفزعة
ومن خواصه انه اذا سحق منه قدر اربع شعيرات
وسقي من به الاستسقاء الماي اسهل الماء من ساعته
وابراه وفي لسان ابن جلدان هو يقوي البصر بحلاوة
ويحفظ نور الناظر ويزيل الغشي والظلمه وقيمته
وزنه المثقال نصف دينار ومن الجوهريين
من يجعل اصناف البنفش خمسة ويجعل الجادي
في المرتبه الخامسه منها الاخيره وذلك بعد
الاسبادنستي ويجعل ثمنه على نصف ثمن الاسبادستي
وقد تقدم ذكر الاسبادنستي الباب الثامن
في الماس علة تكونه في معدنه قال بليوس

الماس

الماس حجر ذهبي اشبه بالاحجار المذابه انه ليس من
الاحجار تسحقه كما تسحق الاحجار بعضها بعضا
فلذلك شبهته بالاجساد ولا يفسده شيء من الاجساد
غير الابار وقلت انه حجر ذهبي واقول ان الماس
انما كان في معدنه في ابتداء خلقته ليكون ذهبا
وذلك ان الماء كان في خلقته في معدنه فلما سخنته
الحرارة تبين الماس الجزء الذي سخنته حدا فصار
حجرا فلما كثرت عليه الحراره عرض فيه غلظ فصار
فيه لزوجه لغلظه وصار اشبه شيء بالزيبق وتوازن
فيما بين رطوبة المعدن ويبسه بلطافة الطباخ ملح
فنشفه الماء والريح فغلظ واشتدت عليه الحرارة
فقوي الملح على سفل الحر واليبس واشتدت يبوسته
فظهر وبطن من على وجه الارض ... الذي هو على

35

وجهه يشبه الزيبق فانعقدت حجرا فافرط اليبس عليه
وانما انعقدت ليكون ذهبا فابعده عن الذهبيه انعقاده
باليبس والملوحه فلما انعقد باللين ولم يفرط عليه
اليبس والحلاوه مكان الملوحه لكان ذهبا لكنه
لما انعقد وفيه ملوحه وشدة يبس نقص عن كيان
الذهبيه وصار حجرا اجلد من كل الاحجار كلها بملوحه
طبيعته وشدة يبسه وانما صار بكثرة الملوحه فيه
الملوحه واليبس حسب وانما صار لا يفسده شيء غير
الابار لانه ذهبي كما ان الابار يفسد الذهب ويسحقه
وانما يسحق الابار الذهب لكثرة تدبيه وذلك لاجتماع
الكبريت الذي في الابار مع ملوحه الماس ولان
الملح الذي في الماس اذا احس برايحة الكبريت يبست
واسحق وانما صار الماس ابيض لانعقاده برطوبه

ودفع رطوبه المعدن عنه وهج النار فصار كذلك ابيض
فهذه علة كون الماس ابيض وفي معدنه الذي يتكون
فيه ويخرج منه كما يخرج الياقوت فهو حصبا معدن
الياقوت ويوجد مع الياقوت اذا اخرجه السيول
والرياح من معدنه ونحن قد بيناه فيما سلف على هذا
اكثر النقل فيه وفي لسان يوحنا بن ماسويه ان
الماس يوجد بواد ببلاد الهند لا يصل الى اسفله
احد من الناس والماس في اسفله حجاره منثوره ما
بين الخردله الى الشعيرة يعمد الى اللحم الطري فيلقى
في ذلك الوادي والنسور تنظر اليه فتهوي خلفه
فتحركه في الارض لتاكله فيلتزق به الماس ثم يتكاثر
عليه وتقتتل فتطير به فيسقط الماس ويلتقط وهذه
النسور معروفة [illegible]

32

حسنة المعقرب ومنوما كان شديد الحمرة ويليه المغلس
وهو اقل حمرة ويليه الاقاري وهو كلون الرمان واكثر
شبها به لان اقانا الفارسيه الرمان ثم السارلي وهو
اقل لونا في الحمرة من الاقاري ثم الاصفر وهو اقربها
شبها بالياقوت وما قرب من شبه الاصفر قيمته
وثمنه قيمة البلخش الجيد في الجمله غالبا على النصف
النصف من قيمة الياقوت ه الباب
السادس في البنفش قد ذكرنا ان تكونه وتكون
البلخش واحد وكذلك معدناهما فاما اصناف البنفش
فهي اربعة ماذني وهو احمر اللون مفتوح صاف
شفاف وهو اعلا انواعه وسالت بعض مشايخ الجوهريين
في سبب تسميه هذا النوع بهذا الاسم قال ان هذا الحجر
شديد الشبه بجيد الياقوت [illegible] اذا قوم بدون قيمة

الياقوت كان يقول بلسان حاله ماذني حتى اقوم
بدون قيمة الياقوت واحمر توى الحمرة البنفش الرطب
والبهيجي وهو احمر اسود وتعلوه حمرة بنفسجيه مبطنه
بزرقه خفيفه والاسبادنشت وهو احمر مفتوح اللون
وجميعه قريب الشبه من البلخش الا انه اكمد لونا ه
خواصه ومنافعه من خواص الاسبادنشت ورحله من
اصناف البنفش يقطع الرعاف بالتعليق من خارج ولا اعلم
لشي من بقيه انواعه خاصيه ه قيمته وثمنه قيمة
البنفش على الربع من ثمن البلخش واقل من ذلك على قدر
صبغه ومائيته واختلاف انواعه والماذني وهو
اعلاه يسوا بدينارين المثقال والاحمر على نصف ثمن
الماذني والاسبادنشت على نصف قيمة الاحمر والبهيجي
على نصف قيمة الاسبادنشت الباب السابع

33

في البجادي ه علة تكونه في معدنه قد ذكرنا
هذا فيما سلف في الباب الخامس معدنه الذي
يتكون فيه يوجد البجادي حيث يوجد الياقوت
في الجزيرة التي وراء جزيره سرنديب بالجبل المعروف
بجبل الراعون المتقدم الذكر وقد ظهرت له معادن
ناحيه بجادي الموجود منه فيها اجود من الهندي
جيده ورديه البجادي حجر فيه حمرة لا شعاع لها
الا الاقل منه وما كان له شعاع فهو شبيه بالياقوت
ومنه حجر اقل حرارة ويبسا من الياقوت واذا اخرج
الحجر من معدنه وجد مظلما ليس له شفوف فاذا
قطعه الصناع خرج لونه ويظهر حسنه واثار ضوءه
وصار له بريق واجوده ما اشتدت حمرته واثارت
برقته وهو لا يضي الا [illegible] على البطاين ثم يحفر

اسفله ويعبا الشاد منه فان الشديد الرطوبه منه
النقي يضي واسفله مسطح غير مقعر الا ان ذلك يقطع
قليلا فاذا وهن من الاحجار حجر يشبه البجادي وهو
الماذنج وهو حجر احمر شديد الحمرة الا انه مايل الى
السواد وهو ارخى من البجادي يحتاج لشدة ظلمته
الى تقعير الحفر في اسفله حتى يبرق والا لم يظهر ماوه
واجوده اكثر ماء واقله حاجة الى كثرة الحفر وهو
دون البجادي في الثمن وينفصل عن البجادي برطوبه
البجادي وكثرة الماء وانه لا يعلق بشي من هباء الارض
خواص البجادي في ذاته من خواص حجر البجادي
انه ان مسح بشعر الراس واللحية ثم وضع على الارض
لصق بشي من ورق التبن وغير ذلك ه ومن خواصه
انه اذا اراته الحيوانات الاناث كلها من ناطق او غير

30

ابتدا ليكون زمردا فقصر عن ذلك بسبب الاعراض الداخلة
عليه من ضعف الطباخ ونقص الحرارة فلان جسمه ونقص
لونه وكان منه الزبرجد معدنه الذي يتكون
فيه الزبرجد يكون في معدن الزمرد ويوجد
منه الا انه قليل اقل وجودا من الزمرد واما في هذا
التاريخ الذي وضعت فيه هذا الكتاب وهو عام
اربعين وستماية فانه لا يوجد في المعدن اصلا وانما
الموجود منه في ايدي الناس على قلته فصوص تستخرج
بالنبش من الاثار القديمة الذي بثغر الاسكندرية
حرسها الله وانها من بقايا كنوز الاسكندرية واخبرني
من نبش عليها بثغر الاسكندرية من الجوهريين ومن
المطالبين معا انه استخرجها من المواضع المذكورة
واراني بعضهم منها فصوصا قال وكنت اجد الفص
وعليه

وعليه قشرة بنفسجية قد سترت لونه فاذا جلي خرج
في غاية الصفا الجوهري حسن المائية ورايت عند
هذا المخبر فصا رايته بحجم من درهم لا يكاد البصر ان
يقلع عنه ولا النظر ان يشبع منه لرقة مائيته وحسن
خضرته وصفايه ذكر لي انه استخرجه بالنبش من بعض
المواضع المذكورة بثغر الاسكندرية جيده ورديه
الزبرجد منه اخضر مغلوق اللون ومنه اخضر مفتوح
اللون ومنه اخضر معتدل الخضرة حسن المائية رقيق
المستشف ينفذ البصر بسرعة وهذا اجود انواعه
خواصه ومنافعه ليس في الزبرجد شي من خواص الزمرد
المذكورة قبل ولا منافعه ولا فيه خاصية سوى حسن
مستشف خضرته وجمالها وان ادمان النظر اليه
يجلو البصر ويقويه قيمته ومثقال زنة نصف درهم

31

من خالصه بدينار الباب الخامس
في البلخش علة تكونه في معدنه قال مصنف
الكتاب البلخش والبنفش والبجادي ثلاثتها من اشباه
اليواقيت كما كان الزبرجد والماس من اشباه الزمرد
واصل تكون اشباه الزمرد اليواقيت الثلاثة المذكورة
واحد ويوجد بمواضع قريبة بعضها من بعض كلها
بلينوس في علة تكونها ان الحجارة الحمر والحمرية مثل العقيق
والبجادي انما انعقدت كلها لتكون ياقوتا فاعترضها
الياقوتية كثرة الرطوبة وقلتها وكثرة اليبس وقلته
وقلة الاعتدال فلم تكن ياقوتا وصارت حجارة حمرا
وحمرية تذوب في النار كما لا يذوب الياقوت ونقع عليها
الياقوت تشاكلها ووضعت عليها الاسما المختلفة
لاختلافها وكل هذه الحجارة انما ابتديت لتكون

ياقوتا فلما عرضت فيها الاعراض لم تتم نسبتها كلها الى
الياقوت كنسبة الياقوت معدنه الذي يتكون فيه
يتكون من معدن ببلخشان والعجم يقولون بلخشاذ
بذال معجمة وهي قاعدة من قواعد معدن النيل مما
يلي تخوم الصين لها اقليم كبير فيه معدن هذا الحجر
واخبرني بعض من وصل الى هذا المعدن ببلخشان
من تجار الفرس ووقف على استخراجه من معدنه
انه راى منه حجرا في المعدن في باطنه ما لم يكمل
طبخه وانعقاده لعدم الحجر يستعمل جيده ورديه
البلخش ثلاثة انواع احمر ويسمى للعقرب واخضر زبرجدي
واصفر واجوده الاحمر وليس لجميعه شي من خواص الياقوت
ومنافعه وانما فضيلته شبيهة به في الصبغ والمائية
والشعاع لا غيره واخبرني بعض الجوهريين ان اصنافه

28

وهو جامع اوصاف الزمرد كلها الظاهرة من اللون
والصفا والرخاوة وخفة الوزن حتى لا يكاد يفرق بينه
وبين الزمرد الا المميز المبرز في نقد الجوهر وخاصيته
التي تفضل بها عن الزمرد انه اذا ركب على البطانة
نقص ماوه وصار الى السواد والصفرة فبان حينئذ
من الزمرد لان من خاصية الزمرد ما ذكرناه من انه
اذا ركب على البطانة زاد حسنه وماوه من اي نوع من
انواع الزمرد كان ومن اشباه الزمرد اليشب واليشم
الاخضر والزبرجد والياقوت الاخضر ذكر
منفعة ... اعلم ان جميع المنافع والخواص المذكورة
فيما سلف انما هي للزمرد الذبابي دون ساير انواعه
وبخواصه يتغالون في ثمنه وقيمة الزمرد الذبابي الخالص
الحجر الذي زنته درهم قيمته اربعة دنانير القيراط

وتتضاعف قيمته بحسب كبره وتنقص بحسب صغره
مع باقي الاوصاف المذكورة قبل في الجودة وضدها الا
ان نقصه في الثمن اقل من نقص غيره من الاحجار بسبب شرف
جوهره وعظم منافعه وكونه جميع الخواص والمنافع
موجودة في الحجر الكبير منه والصغير والمعوج والمستقيم
فاما بقية اصناف الزمرد الثلاثة سوى الذبابي
فانها لا قيمة لها يعتد بها لعدم خواص المنافع الموجودة
في الذبابي منها واخبرني القاضي الخطيب معين
الدين مبشر امين السلطان على معدن الزمرد بالديار يار
المصرية قال وجدت بمعدن الزمرد المعروف
بوادي الشاه قصبة وهو اكبر معادنه من قوص
وعيذاب قصبة زمرد ذبابي وقع الضرب عليها في الحجر
من يد الصانع في المعدن فانكسرت وهي مغيبة في الطلق

و ...

29

فجمعت كسورها ووزنت فكانت ثمانية وثمانين درهما
ووجدت بعد ذلك في المكان نفسه اسفل من مكان
هذه القصبة التي انكسرت بقامة قصبة ذبابي وزنها
ستة دراهم حملتها الى الخزانة المعمورة الصالحية اعلا
الله امر مالكها قال واشتريت بقوص في مدة
السلطان الاعظم الكامل قدس الله روحه من رجل
من التجار قصبة زمرد وجدها في المعدن المتاخم
لبلده من نوع الزمرد الريحاني فكان وزنها بعد
العمل من المنهدم بالحك والجلا اثنى عشر مثقالا
قال وكنت اشتريتها من التاجر وحملتها الى السلطان
الكامل قدس الله روحه وهو بدمشق فقومت
بدمشق بثلاثين الف درهم ومعدن الزمرد من
قوص على مسيرة عشرة ايام وفيه عدة معادن لم
يعمل الى الان فيها شي وهي مكامن وقوع غازي ووادي
السهر وهو معدن كبير ووادي الشاه سمي بذلك لانه وجد
فيه تمثال شاه من نحاس واخبرني هذا الشيخ المذكور
انهم يجدون في هذه المعادن رطوبة مجتمعة تشبه
الزجاج وارانى مما حمل من هذا المعدن حجارة سودا
رصاصية تقبل الجلا ذكر انها زمرد اسود وحجارة اخر
بين البياض والصفرة هشة كانها اول تكون الزمرد
وقد ظهر بمصر على يد هذا الرجل المذكور بمسيرة بعض
يوم من مصر نفسها بموضع يسمى طرا من معدن ياقوت
احمر اراني منه هذا الرجل فصوصا حمرا صغارا كالحب دل
واكبر قليلا

الباب الرابع في

الزبرجد هو علة تكونه في معدنه تكون الزبرجد
في معدنه على نحو ما ذكرناه من تكون الزمرد وكان

26

ذكر عيوب الزمرذ ٭ من اكبر عيوب الزمرذ
الذبابي اختلاف الصبغ حتى يكون موضع منه مخالفا
للون موضع آخر ٭ ومن عيوبه عدم الاستواء في الشكل
وهذا عام له ولليـاقوت ولكل حجر يستشف ثمين او
غير ثمين ٭ ومن عيوبه التشعير وهو من لوازمه لا يكاد
يخلو منه وهو شعر شقوق خفية تظهر فيه خواص
الزمرد الذبابي في ذاته خاصية الزمرد الذبابي في نفسه
وهي التي انفرد بها عن سائر الاحجار وبها يمتحن الخالص
منه من غير الخالص ان الافاعي اذا نظرت اليه ووقعت
ابصارها عليه تفقأت على المكان قال احمد
التيفاشي قد كنت اقف على ذكر هذه الخاصية في الزمرد
في كتاب الاحجار ثم جربتها بنفسي فوجدتها صحيحة وذلك
انه كان وقع الي فص زمرد ذبابي خالص اردت امتحانه
على عيو

على عيون الافاعي فاستاجرت حاويا على صيد افعى وصادها
فجعلتها في طشت واخذت قطعة شمع والصقتها في
راس سهم ثم الصقت فيها الفص وقربته من عيني الافعى
فوثبت ولا نحو السهم فكانت لها حركة وقوة قوية تروم
بها الخروج من الطشت فلما قرب الزمرد من عينيها سمعت
فرقعة خفية كمر بتقيل صوابة على ظفر ثم رايت عيني
الافعى قد برزتا على وجهها بروزا ظاهرا وبقيت حائرة
تدور في الطشت لا تقصد مخرجا ولا تدري حيث تتوجه وسكنت
اكثر حركتها وانقطعت قوتها بالجملة ٭ ومن خواص الزمرد
الرخاوة وتحليل الاجزاء ٭ ومن خواصه خفة الوزن
وهو تابع للرخاوة والتحليل ٭ ومن خواصه زيادة
الخضرة والماء اذا ركب على البطانة ٭ ومن خواصه شدة
الملاسة والصفاء والنعومة واخص صفات الزمردية

27

الخفة والتشعير فانه لا يكاد يخلو من التشعير كما ذكرنا واما
الخفة فهي ذاتية له في اصل تكونه وقد ذكرنا العلة
في ذلك في ذكر تكونه والزمرد يتحجل على النار ويتكلس
فيها ولا يثبت عليها كما يثبت الياقوت وسبب ذلك
ما ذكرنا من رخاوته وتحليل اجزائه خواص الزمرد
في منافعه من خواصه انه من ادام النظر اليه اذهب
عن بصره الكلال ٭ ومن خواصه انه من تقلد بحجر منه
او تختم به دفع عنه الداء المعروف بالصرع اذا كان لبس
قبل حدوث الداء ٭ ومن اجل هذا كانت الحكماء تامر
الملوك ان تعلقه في اعناق اولادها عند ولادتهم
ليدفع عنهم داء الصرع ٭ ومن خواصه ان من حل
منه ثمان شعيرات وسقاه شارب السم قبل ان يعمل
السم فيه خلص نفسه من الموت ولم يسقط شعره ولم

ينسلخ جلده وكان شفاء ٭ ومن خواصه انه ينفع من
نفث الدم واسهال الدم ومن العلة المعروفة بالذوسنطاريا
تعليقا على المعدة والكبد من خارج شاهدت بنفسي من فعل
ذلك بالمغرب فبرأ ثم وصفته لمن كانت عليه ذوسنطاريا
بالمشرق وعجز الاطباء عن مداواتها واشرف على الهلاك
تعلقته فكان برؤه به باذن الله ٭ ومن خواصه
النفع من وجع المعدة ايضا بالتعليق عليها من خارج ٭
ومن خواصه ان باقي الحيوانات جميعها ذوات السموم لا تقرب
حامله ٭ ومن خواصه ان جميع اصنافه كلها تعلق
على العضد للرقية وللتعويذ وعلى الفخذ لسرعة الولادة
مجرب ذلك كله ٭ وقال المعلم ارسطو ان الزمرد اشد
بردا ويبسا من الدهنج انه صيني يتجسد في الارض وفي
اشباه الزمرد شيء يسمى الماش يخرج مع الزمرد من معادنه

24

وانما خفي الزمرد وخف لان اخفاء الينبوع الذي فيه
وانما انحلت بالاعتدال ولين الطباخ فلما انحلت اليبوسة
فيه ولم ترجع منقبضة كقبض الياقوت فيصير ثقيلا واشتد اجزاؤه
اجزاؤه لكنها انعقدت بلين النار وطيب الطبائع انعقدت
بالاعتدال وكما تم انعقاده كملت عليه الحرارة
بحرها ويبسها فيبست اجزاؤه وتكاثفت منافذه
فحجبت عن النار ان تؤثر فيها فهذه علة
الزمرد والزبرجد وتكونهما في معدنيهما كما بلينوس
وقد امكن الياقوت ان يكون زمردا او الزمرد ياقوتا كما
امكن الفضة ان تكون ذهبا والنحاس فضة بانقلاب
بعضها الى بعض اذا كان اصلها من شيء واحد وانما اعترضت
فيها فاختلفت جنسها كما تبينا في الاجساد الذائبة
انها تنقلب الى لون حتى تصير الى جوهرها الذي ابتدأت

منه كذلك الاحجار على مثال الاجساد

ذكر معدنه الذي يتولد فيه موضع
الزمرد الذي يؤتى به من التخوم من بلاد مصر والسودان
خلف اسوان يوجد هنالك ممتدا كالحبشة فيه معادن
تحفر فتخرج منها الزمرد قطعا صغارا كالحصباء
منبثة في تراب المعدن واخبرني رئيس المعدنين
بمصر المكلف من قبل السلطان بهذا المعدن ان اول
ما يظهر من معدن الزمرد شيء يسمونه الطلق ومنه
حجارة سود اذا احمي عليها في النار خرجت مرقشيتا
ذهبية قال ثم يحفر فيجد طلقا يشابه الزمرد في
تربة حمراء لينة مشتملة عليه وربما اصيب لعرق منه
متصلا فيقطع وهو اجوده واما صغيره فانه يصاب
في التراب يجمع من نخل وذلك انهم ينخلون التراب ثم

25

يؤخذ حلاله فيغسل كما يغسل تراب الفضة فيوجد الحجر
بعد الحجر يوجد بعضه على تربة سوداء كالكحل الا انه
صغير كالخردل او اكبر قليلا اذا ثبت هذا المعدني
وما وجد منه في التراب فهو الفص وما قطع منه من
العرق فهو القصب في اصطلاح الجوهريين والمعدنيين
هكذا وهو اعتقه واجوده واخلصه كما ذكرناه

ذكر جيده ورديه اصناف الزمرد اربعة
الذبابي والريحاني والسلقي والصابوني فاعلاه
واغلاه وافضله في جميع الخواص الموجودة في الزمرد
هو الذبابي وهو اخضر مغلوق اللون جدا لا يشبه
خضرته شيء اخضر من الالوان له حسن الصبغ جيد المائية
وانما قيل سمي الذبابي لشبه لونه بالخضرة التي تكون في
الكبار من الذباب الريحان الربعي الموجود في البساتين لاني

مغاني الموجود في البيوت وهو احسن ما يكون من الخضرة
سبحان الله اللون بعينه موجود في ذباب البيوت
واما بقية الاصناف المذكورة من الزمرد غير الذبابي
فانها نازلة مقصرة عن جميع الخواص الموجودة في الذبابي
والريحاني مفتوح اللون كلون ورق الريحان ودونه
السلقي كلون السلق ودونه الصابوني كلون الصابون
ولا يتم له يعتد بها واحسن اصنافه الذي يضرب الى
البياض مع كمد ويسمى العربي وهو يوجد في برية العرب
في ارض الحجاز وقد قلنا ان اجود الزمرد الذبابي ومنه
اشد صفاء في نوعه حتى السنوب خضرته شابه من صفرته
او ميل الى السواد او غير ذلك من الالوان مع شدة
الشعاع فان انضاف الى ذلك كبر الجرم واستواء
القصبة وعدم الاعوجاج كان الغاية والنهاية وكان ثمنه اعلا

22

ولا يزيد بعض الزيادة متقاربة هي على ما اصف الحجر
الذي زنته نصف درهم قيمته ستة مثاقيل من الذهب
الخالص تكون زنته قيراطاً منه بعشرة دراهم من الفضة
النقرة الخالصة لها من الذهب العين الخالص نصف
وربع مثقال الحجر الذي زنته درهم قيمته ستة عشر
ديناراً زنة كل قيراط منه بدينارين الحجر الذي زنته
مثقال قيمته بدينارين ونصف للقيراط الحجر الذي
زنته مثقال وثلث قيمته ثلاثة دنانير للقيراط الحجر
الذي زنته مثقالان قيمته ما بين اربعة دنانير القيراط
الى ثلثه وتصف على قدر لونه وزيادة ما بينته وتزيد
ثمنه بحسب زيادة لونه وما بينته وكبر جرمه وصغره
وربما بلغ ما زنته مثقال من الذهب اذا كان نهاية
نهاية في الصبغ والمائية والشعاع مصموعاً قد نقص

منه بالصبغة والحل والغل والجلاء كثير من جهة فاما
الازرق والذهبي منه فقيمتها واجل قيمة كل درهم
منها اربعة دنانير واما الاصفر فقيمة زنة كل درهم منه
ديناران واما الابيض فقيمته على النصف من ثمن الاصفر
ويختلف ذلك كله بالزيادة والنقصان في الصبغ والماء
كما ذكرناه آنفاً الا ان اختلافه في ذلك راجع الى
الاصل الذي بيناه وقرات في بعض كتب التواريخ
ان سبب نكبة هشام ابن عبد الملك لخالد بن عبد الله
ابن اسد انه رفع اليه انه اشترى حجر ياقوت احمر يغمض
عليه باليد فيفضل عند القبضة من جهة الكف بثلثين
الف دينار فطلبه منه فانكره وكتم الحجر وكتمه من بعض
خطاياه وذكر ابو عبد الله الغلابي في كتاب المؤلف
في اجواد قريش في رسم هشام ابن عبد الملك ان يوسف

23

بن عمر بعث الى هشام المذكور بياقوتة حمراء يخرج طرفاها
من كفه وحبة لؤلؤ كاعظم ما يكون من الحب
وان الياقوتة كانت للرائعة جارية خالد بن عبد الله
وانه اشتراها منها بثلاثة وسبعين الف دينار ۞
وذكر بن بختيشوع ان مصادقة زمان جعفر
البرمكي وهبته جاماً كان بين يديها تاكل فيه وكان
معه معلقة كلاهما من الياقوت الاصفر فباعهما فكانا
سبب غناه عمره كله ۞ الباب الثالث
في الزمرذ ۞ لغة الزمرذ بضم الزاي والميم والراء
المشددة وبذال المعجمة كالذي تكلمت به العرب
وقال التاشي في كتابه في اللغة ان الزبرجد يعرب
الزمرذ وليس كذلك بل الزبرجد نوع آخر من الحجارة
ياتي ذكره بعد هذا الباب ان شاء الله تعالى ۞

علة تكونه في معدنه قال بليناس الزمرذ هو
الياقوت وانما ابتدا ان يعقد ياقوتاً في جميع اجزائه وكان
لونه احمر ولشدة الحمرة عليه بعضها على بعض عرض له
السواد فصار اسمانجونياً لفعل اليبس وعلة الاسمانجوني
ان ارتفع ما صفا من الحمرة على اعلاه اصفر وباطنه
اسمانجوني واشتدت عليه الحرارة فطبخته ومزجت
اللونين جميعاً لون ظاهره بلون باطنه فتولدت
الخضرة بينهما فصار لونه اخضر فسمي زبرجداً وانما
كان اصله ياقوتاً لان الياقوت حجر ذهبي وهو اصل
الحجارة كما ان الذهب راس الاجساد المنطرقة وكان
الياقوت ايضاً لا تذيبه النار ولا يبرده الحديد لانه
يبسه ظاهر على اعلاه من شدة تكاثف اجزائه بنسبة
بعضها في بعض فظهر يبسه على وجهه فلم يعلق به الحديد

20

تزداد وقد ذكر ارسطو في كتابه في الاحجار ان
الياقوت الاحمر اذا نفخ عليه في النار ازداد حسنا وحمرة
واذا كان فيه نكتة شديدة الحمرة ونفخ عليه في النار
انبسطت في الحجر تشعشع من تلك الحمرة حسنته وان
كان فيه نكتة سوداء انقص سوادها وهو حجر يزداد
حسنا وصفاء عند النفخ عليه في النار واذا كان الحجر
احمر وحمي في النار فذهبت حمرته فليس بياقوت تلك
احجار الاشباه او هو مصنوع مدلس وقد رايت بسوق
الجوهريين بالقاهرة المعزية حماها الله تعالى
حجارة تباع على انها ياقوت ازرق واصفر وهي مصنوعة
مدلسة كان اصلها ياقوتا ابيض ومن خواصه انه
لا تعمل فيه المبارد والحديد ولا يلصق شيء من حشمه
من جميع الوانه احمره واصفره وسماويه وان من خواصه

يقطع

تقطع الحجارة المستفدة غير الماس ولا في جميع هذه الخواص
آبد على الوانه في القوة خواص الياقوت في منافعه
من خواص الياقوت ما ذكره ارسطو وذلك انه من تقلد
بحجر او تختم به من اجناس اليواقيت الذي وضعنا وكان
في بلد قد وقع فيه الطاعون ونبل في اعين الناس
سهل عليه قضاء الحوائج وتيسرت عليه من اسباب
المعاش امور صعبة ومن خواصه تقوية قلب لابسه
وتشجيعه والهيبة له في قلوب الناس واجلاله ومن
خواصه انه ينفع من الخفقان والوسواس بالتعليق
ومن خواصه ان الصاعقة لا تقع على من تختم
به او علق عليه ومن خواصه انه يقطع من جمود الدم
اذا علق ومن خواصه انه ينفع من نزف الدم اذا علق
ومن خواصه ما اخبرني به جوهري شريف معروف

21

بالخبرة والذكا في هذا الفن دخل الى الهند ونال شيئا
كثيرا من علم الاحجار ان اهل الهند يقولون ان من كان
معه حجر ياقوت جذب نفوسا قوية على طبعها الى طبعه
وقوته بحيث انه اذا لم يكن معه ذلك الحجر على شرط ان يفعل
ذلك على جهة الخبرة والامتحان بل يكون ذلك بغير تفقد
ولا تعمد ومن خواص الاصفر منه انه يمنع الاحتلام
والياقوت الخالص اشباه تقارب الوان هذه اليواقيت
التي قدمنا ذكرها غير انها ليس لها نصاعة الالوان
الاصلية ولا خصوصيتها ومحنة اشباه اليواقيت
باجمعها ان تحك بالياقوت الاحمر فانه يجرحها كلها
ولا تجرحه وليس منها ما يقوم على النار ولا يحترق
والياقوت الاحمر يثبت في النار على حسنه وصبغه بل
يزيد حسنا كما ذكرنا ان اصناف اليواقيت

واثمانها اعلم ان جميع اثمان الاحجار وقيمتها تختلف
بحسب امرين احدهما في ذلك الحجر والاخر في الاسباب
الخارجة عنه فاما الذي في ذاته فامران احدهما جودة
جودة الحجر ورداءته والثاني كبره وصغره فاما الاسباب
الخارجة عنه فمنها نفاق السوق وقوة الشهوة ومنها
اختلاف البقاع في البعد والقرب من معادن الحجر
ونحن نضع قيمة الاحجار الذي نذكر في هذا الكتاب
بحسب اعتبار سوقها في موضعين وهما بغداد ومصر
اذ كانا كالوسط الذي منه سائر الاطراف اليه راجع
والحجر الجيد الخالص الكامل الصفات في نوعه يختلف
بحسب كبره وصغره فقط كلما عظم جرم الحجر تضاعفت
قيمته وكلما صغر نقصت قيمته حسبما نبينه آنفا فالياقوت
الاحمر الخالص البهرماني قيمته الوسطى المتعارفة سعرا

18

قريب من لون من لون الورس في الضعف والبهرمان
اسم العصفر وبه سمي هذا الصنف من الياقوت واثمن
الياقوت الذي في لون الجمرة وهو البهرماني واثمن كل
واحد من بقية اصنافها اشبعها صبغا وارقها
مستشفا واشدها شعاعا واسلمها من العيوب التي
نذكرها فيما بعد فاما الياقوت الاصفر فمنه الرقيق وهو
قليل الصفرة كثير الماء ساطع الشعاع والخلوقي
وهو اشبع صفرة من الرقيق والجلناري وهو اشبع من
الخلوقي واشدها شعاعا واكثرها ماء وهو اجود
اصناف الياقوت الاصفر والاسمانجوني فمنه الازرق
واللازوردي والنيلي والكحلي وهو اشبع من النيلي
ويسمى الذبابي واما الابيض فمنه المهاني وهو اشدها
بياضا واكثرها ماء واقواها شعاعا واصلبها حجرا

وهو ادونها وثمنه ارخص اثمان جميع اليواقيت كلها
ذكر خالص الياقوت ن اجود الياقوت الاحمر
البهرماني والرماني والوردي النير المشرق اللون الشفاف
الذي ينفذه البصر سريعا السالم من العيوب
عيوب الياقوت الشعرة والسوس والشعرة
شبه تشقيق يرى فيه والسوس خروق توجد في باطنه
يعلوها شي من ترابية المعدن وربما وجد في تلك
الخروق دود حي يتحرك فاذا اخرجت الدودة منها الى
الهواء ماتت واخبرني من راى ذلك من الثقات ن
عيوب الوانه اردى الالوان الاحمر اللون الوردي
يضرب الى البياض والسماقي الذي يضرب الى لون الرماد
ويسمى الستوري وكذلك الذي يسمى الذبيبي واردى
الوان الياقوت الاصفر ما بعض لونه يضرب الى البياض

19

واردى اصناف جميعه في الجملة قبح الشكل والتغيار والتطرابق
والنقور ويسمى السوس طبع الياقوت وقوته قوة الياقوت
على قدر المعادن التي يتكون فيها وعلى قدر المعادن
صبغته والوانه فالاحمر منه حار يابس والاصفر او بها
الى الاحمر وفيه فضل حر ولذلك اصفر والاسمانجوني
ابيض واليبس والابيض ابرد اليواقيت وارطبها ن
خواص الياقوت في ذاته ن من خواص الياقوت
انه يقطع كل الحجارة شبيها بقطع الماس وليس يقطعه
شي غير الماس ولذلك ان ركبت منه قطعة على
سفت حديد فانه يثقب كما يثقب الحشب ن ومن خواص
الياقوت انه لا ينجلي على خشب القشر الذي ينجلي عليه كل
شي الا الياقوت فانه لا ينجلي على شي الا على صفيحة نحاس
كسر الجزع اليماني ويحرق حتى يصير كالنورة ثم يسحق

بالماء حتى يصير كانه الغرا ثم يجلى به على وجه الصفيحة
النحاس حجر الياقوت فيتجلى حتى يصير اشد الجواهر لمعانا
ومن خواص الياقوت الشعاع فانه ليس لشي من الشفاف
شعاع مثله ن ومن خواصه الثقل فانه اثقل المساويه
لمقداره في العظم ن ومن خواصه صبره على النار لا يتكلس
غيره من الحجارة المثمنة كالزمرد وغيره من الحجارة
ومن خواصه انه يقبل البروده بسرعة اذا اخرج من النار
بخلاف غيره من ساير الاحجار فاما اصباغ الياقوت فانما
نثبت منها على النار الحمرة فقط واما غيرها من ساير
الوانه كالصفرة والاسمانجونية والسواد فانها تنسلخ
كلها بالنار ويبقى حجرا ابيض ويتكلس وتنقب ان اوقدت
عليه النار والصفرة ابعدها انسلاخا والسودة اقلها
ثباتا على النار واما الحمرة وحدها فانها لا ينفك

16

وتراه لبقر وغيره من الحيوانات بحيث اذا ابتلعته
عمدت الى اصل شجرة فالتوت عليه واستدت بيكسر
في بطنها ما تبتلعه وتندق عظامه فيهضم لهلاك
فلاجل ذلك ايضا لا يستطاع سلوك هذا الجبل ولا
الوصول الى ما فيه من عجايب الاحجار ومن الياقوت
ما يوجد في هذه الجزيرة في الحما والوحل يطلب منه
الان اجوده ما حملته السيول من الجبل المذكور وعلامة
الجودة فيه كثرة الماء والصبغ والشعاع والياقوت الاحمر
يخرج من جميع هذه المواضع بعضه احمر وبعضه
احمر شديد ما والشديد الحمرة يكون فيه سواد يغلب
عليه ولا تتبين الحمرة الا عرفا في ذلك السواد وقد يكون
في الحجر موضع خال تكون فيه ريح وربما كان ايضا فيه الما
وربما كان فيه طين ويرى كل ذلك من خارجه فيوضع عليه
السراج

المثقب حتى تذهب ذلك العيب ويخرج الماء والريح
والطين منه وكذلك يكون فيه خرق مثل السوس في
الخشب تسمى المريم يوضع المثقب عليه من ذلك الجانب حتى
ينقطع عنه وان ترك في الحجر من عيوبه شي لم يخرج
انصدع واكثر ما يوجد الغالب عليه السواد فيخرج لونه
بسبب المعالجة بالنار وصفة عملهم فيه انهم
ياخذون حصى من حصى تلك الارض التي يوجد فيها
ذلك الياقوت فيسحق ويحل بالماء ويسحق حتى يلزم بعضه
بعضا ثم يطلى على الحجر حتى يغيب فيه ثم يوضع على الجمر
وحوله حجارة ويلقى عليه الحطب الجزل وينفخ عليه ويديم
النفخ والقا الحطب على مقدار السواد يعرفونه بالدربة
المزاولة وينفخ عليه من بين ساعة واحدة الى عشرين
يوما وليلة ولا يقلع عنه والتعاهد بالحطب ليلا ونهارا

17

على قدر ما فيه من السواد فان اخرجوه عند تعاهدهم
اياه وقد ذهب سواده وصار الى لون من الالوان
كان بها ما كان غير السواد لم يعيدوه الى النار لانه بعد
خروج السواد منه لا يزيد لونه ولا يستغرق هذا اعلاجه
سرنديب وقد يعالج بعضه ببلاد اليمن وذلك قليل
قال [illegible] حكى ماسويه ان اخذت قطعة من
الياقوت الاحمر عالية وجدتها مثاقيل قال
والياقوت الاسمانجوني يجوز ان يكون في بعضه صفرة
فيوضع في الجمر فتذهب صفرته فان جاوز مقدار النار
ذهب اللون الاسمانجوني وصار ابيض والاسمانجوني
والمصفر اذا وضعا في النار ابيضا ولا يتغير عن البياض
قال وتكون الفيلجه من الاسمانجوني اربعين مثقالا
ذكر اصناف اليواقيت وبيان الجيد والردي

منها هو اصول اليواقيت اربعة انواع احمر واصفر واسمانجوني
وابيض والاحمر منه ينقسم الى اربعة اقسام الوردي
وهو احمر لون الورد وهو يتفاضل في شدة الصبغ الى
حد الوردية الجيدة كذلك ويقل صبغه الى ان يقرب من
البياض ثم الخيري وهو احمر مشرب بعد فرقين كلون
الورد الخيري وامهر في خيريته وهو يتفاضل في
قوة الصبغ وضعفه الى ان يقرب من البياض ثم الاحمر
وهو يكون كلون العصفر الشديد الحمرة وهو ايضا
يتفاضل في قوة الصبغ حتى ينتهي الى لون العصفر
الناصع في القوة والى القرب من الوردي في الضعف
ثم البهرماني وهو احمر نقي الحمرة لا يشوبها شايبة
وهو يتفاضل ايضا في قوة الصبغ وضعفه حتى ينتهي
الى لون العصفر الشديد الحمرة الناصع في القوة والى

14

ثم تكاثفت اجزاوه بعضها في بعض وتداخلت فهذه علة تكون
الياقوت، فاما اختلاف الوانه فانه بسبب بقاع الارض
اذا وقع عليها الماء فدام عليها فيتغير الماء بما اتخذ
فيه من جنس الارض وتسخين الشمس له فيجي على قدر الحرارة
وسعتها احمر وربما انعقد اصفر لقلة الحرارة فيه وربما
وربما اعتدل الحر عليه في اللين وللاعتدال فانعقد
ابيض صافيا وربما اشتدت يبوسته فعرض فيه البرد لشدة
اليبس وتباعد الحر عنه فعرض فيه السواد وظهر على
اعلاه لطول الحمرة في باطنه فربما طرحت الحمرة نورها
الى خارج مع ظهور السواد في ظاهره فقام بينهما لون
اسمانجوني وذلك ان صفرة الرطوبة اذا التحمت بغلظ
سواد اليبس قام بينهما اللون الاسمانجوني وكل
ملبوس والياقوت حجر ذهبي وجميع الحجارة غير الاجساد

الذائبة

الذائبة انما انعقدت وابتدات لتكون ياقوتا كما ابتدات
الاجساد الذائبة كلها لتكون ذهبا فعاقتها عن الذهبية
العوارض وكذلك الاحجار انما ابتدات في خلقها لتكون
ياقوتا فعاقتها عن الياقوتية كثرة الرطوبة وقلتها
وكثرة اليبس وقلته فلم تكن ياقوتا وصارت حجارة حمرا
او بيضا او خضرا او صفرا او غير ذلك من الالوان لا تذوب
في النار ويقع عليها الحديد فيصلحها ومنها ما لا يصلحه
الحديد ووضعت عليه انما كثيرة خلاف الياقوت
ذكر معدنه الذي يتكون منه الياقوت يؤتى به
من معدن يقال سحيران من جزيرة خلف جزيرة
سرنديب بنحو من اربعين فرسخا والجزيرة نفسها تكون
اثنين وستين فرسخا في مثلها وفيها جبل عظيم يقال
له جبل الراهون تحدر منه الرياح والسيول الياقوت

15

فيلتقط وهو حصى ارض ذلك الموضع وحصاه وما سقط
من السيول ويقال ان الشمس اذا اشرقت على ذلك الموضع
انبثت له شعاعات كثيرة لوقوع شعاع الشمس على
حصا الياقوت فيسمى ذلك برق الراهون وهذا الجبل
هو الذي اهبط عليه ادم صلوات الله عليه من الجنة
ومنه خرج الى الارض واذا انصب ذلك الحصا اصيب ظاهره
مظلم يميل اكثره الى السواد والغبرة كالحصا الموجود
عندنا في هذه الالوان فاذا استشف في الشمس
شف لونه احمر كان او اصفر او سماويا او غير ذلك
من الوان الياقوت قال احمد التيفاشي رحمه الله
واخبرني من دخل جزيرة سرنديب من التجار ان اهل
ذلك الموضع اذا لم تحدر السيول والرياح من حصى الياقوت
في بعض السنين ما جرت به العادة احتالوا في تحصيله

بالحيلة التي نذكرها وذلك ان الجبل الذي فيه الياقوت
جبل شاهق صعب المسلك لا يمكن الوصول الى اعلاه وفي
اعلاه نسور كثيرة تعشش فيه وتتخذ مساكنها به لخلوه
من الانس فيعمد اهل ذلك الموضع الى حيوان فيسلخون
جلده ويقطعونه قطعا كبارا ويتركونه في سفح جبل الراهون
ويبعدون عنه وهم يرقبونه فتاتي النسور فترفع ذلك
اللحم وتنزل به عند اوكارها فاذا وضعته على الارض
علق به حصا الياقوت ولصق به ثم تاتي نسور اخرى
فتهجم على اللحم لتخطفه فتاخذ بعضها وتطير به في
الجو فيسقط منه الياقوت لثقله فيلتقطه الذين
يرقبونه من الموضع الذي سقط فيه ويذكرون ان في اسفل هذا
الجبل غياضا عظيمة وخنادق عميقة واشجارا شاهقة
تسكن فيها حيات عظام تبتلع الحية منها الانسان

12

بالتسوية يسحق ناعما ويعجن بلبن حليب ويطلى به الحب طليا
ثخينا واجعله في عجين والفه في تنور حتى ينضج فان الحبة
بيضا وان كان قد غير الطلي تاخذ جزا من الصابون
ومثله من الملح الدراني وخذ نوزه واجعل ذلك في انا
زجاج وصب عليه ما عذبا واجعل الحب فيه على نار
فحم لينه وكلما رغا الصابون خذ رغوته وجدد الما
افعل مرارا واغسله بالما القراح فانه يستوي ويبيض
ويرجع الى احسن ما كان عليه صفة اخرى في جلا
اللالي تجعل كافورا مسحوق في خرقة كتان رقيقة وتنزل
الحبة في وسط الكافور بعد ان يدق الكافور ناعما
وربط الخرقة وتوضع في قدح زجاج فيه دهن حب المحلب
او دهن زيبق على جمر ضعيفة قدر ما تعد خمسمائة عدة
ثم يخرج وقد انجلا وان بقيت فيه بقية اعدته الى العمل

فانه ينجلي صفة اخرى يجعل بدل الكافور طباشير
والعمل واحد بدهن حب المحلب والزيبق صفة اخرى
حماض الاترج مقطرا ادخل خمر قوي معطر حلوان ما
كانت صفرته فيه ظاهرة مجرب صفة اخرى نوشادر
قيراط تنكار حبتان ورق حسه تلي ثلاث حبات
يدق الجميع ويجعل في مغرفة جديد ويصب عليه خل
خمر جيد قدر حشوة ويغلى عليه غلوتين فاذا غلى نعما
وضع علته المغرفة في ماء بارد من غيران ينال الدوا
شي فاذا برد القي فيه شيا من ملح ذراني مسحوقا
ثم تاخذ من هذا الدوا في راحتك وتدلك به الحبة
دلكا جيدا ثم تغسلها بالما العذب يخرج كالجوهر
الباب ______ الثاني في الياقوت
لغته من اسمايه الجوهر والكبريت والعسجد في بعض

13

اللغات علة تكوينه في معدنه قال بالينوس في كتابه
الملقب بسر الطبيعة في العلل والمعلولات ان الحجارة
ضروب شتى والالوان مختلفة منها صافية ومنها كدرة
ومنها صافية شديدة ومنها رخوة ومنها ما يذوب في
النار ومنها ما لا يذوب ومنها ما يتكلس ومنها ما لا
يتكلس واصلها كلها الذي نمت منه هو الما والتراب
بالزيادة فيها والنقصان وبقدر المكان الذي
تتوارى فيه وبقدر طلوع الشمس ودوامها عليها في موضعها
وبقدر ما احتجبت عن الشمس ولذلك اعترضت عليها
العوارض من الشدة والرخاوة والطعوم الحريفة
والروايح والالوان والعلة في تكون حجارة الياقوت
هي ان الشمس لما طلعت على الارض سخنها بقوتها وسخن
سطح الارض ما لم يحجب عنها شي واشتدت سخونة المكان

وظهور الشمس عليه فقطرت الشمس رطوبة المكان الذي
اشتدت حرارتها عليه فلما اشتد يبسه لقلة رطوبته
اخذت قوته من الشمس وقوتها حرا ويبسا فانقلب
من طباعه ولونه وطعمه على قدر الرطوبة التي كانت فيه
من كثرتها وقلتها حبست الرطوبة وقامت عليه اخذت
الما ما كان في ذلك المكان من حر الشمس ويبسها وطلعت
عليه الشمس وتحته حجب الضوء به عن ذلك اليبس الذي
فيها بحر الشمس فسخن الما بحرها فتلطف وقوي على
تحليل اليبس الذي قبلته الارض من يبس الشمس المتصل
بها في الما وانحلاله واشتدت عليه السخونة حتى ظهرت
قوة اليبس المفرط فيه فكان منه الحجر المسمى الياقوت
واشتد به يبسه ضاقت عليه مسامه لقبض اليبس له ولشدة
انحلاله وشدة لطافته رجع منعقدا ولسبب اليبس

10

الفائزة النقية اللون الجنبة المايه وهي المصيص الشفافه
الجوهر وهو الاشراق وما كان من الجوهر بهذه الصفة تسمى
في اصطلاح الجوهريين الرطب واذا كان وزن الجوهرة
مثقالا وهي بهذه الصفة فقيمتها ثلثمايه دينار واذا
كانتا جوهرتين الواحدة منهما وزنها مثقال وهي بهذه
الصفة وهما شكل واحد لا يفرق بينهما في الشكل والصورة
كانت قيمتها اكثر من سبعمايه مثقال لاجتماعهما ومناسبتهما
في النظم واذا كان وزن الاثنين دينارا مثقالا وهما بهذه
الصفه كانت قيمتها مايه دينار وان كان وزنها ثلثي
مثقال كانت قيمتها ايضا عشرين دينارا وان كان وزنها
نصف مثقال كانت قيمتها عشرين دينارا وان كان
وزن ثلث مثقال كانت قيمتها خمسة دنانير والجوهر
يحتمل الزيادة في السوم عند الرغبة فيه الا ان العيب

فيه لا يغفر وليس يشترط بعض الثمن لكن معظمه وسبب
ذلك ان المنفعة فيه انما هي بحسب شكله وصورته لا بخاصية
اخرى فيه كخواص غيره من الاحجار والمراد منه انما هو
الزينه والتجمل به فاذا عدم منه جمال الصورة وحسن الشكل
لم يبق له معنى يوجب الغبطة فيه والله اعلم [illegible] ما زاد
على وزن درهمين ولو وجد واحد في الجوهرة الواحدة فانه
يسمى في اصطلاح الجوهريين درا فان نقصت في الوزن
عن درهمين ولو وجد ايضا تسمى جوهرة في اصطلاحهم
وجاهدا بشرط اجتماع سائر الاوصاف الجيدة في الدرة
وان كانت زنتها اكثر الا ان فيها شيا من عيوب الجوهر
المذكورة فيما قبل فانها تسمى ايضا حبا في اصطلاحهم ولا
يعتد بوزنها مع عدم الاوصاف الجيدة فيها والدرة وهي
التي وزنها درهمان وحبه مثلا او حبتان اذا اجتمعت

11

فيها الشرايط المذكورة في الجودة كانت قيمتها سبعمايه
دينار [illegible] وان كانت اثنتين على الصفة المذكورة كانت
قيمتهما الفي دينار الواحدة منها بشرط اجتماعها بالاخرى
عيوب الجوهر [illegible] التضديف وعدم الاستدارة
والصفرة والانبزام وقبح البياض وحصبته وعدم رونقه
وسعة الثقب وصغر الجرم وخفة الوزن الاشيا التي
[illegible] بالجوهر الادهان جميعها والحموضات باسرها
وخاصة الليمون وهيج النار والعرق وزفر الرايحه
والاحتكاك بالاشيا الخشنة في جلا الجوهر
اول ما ننظر من جلا الجوهر ان نعتبره هل تغيره من
اصل معدنه او طارى طرى عليه بالاستعمال وذلك يعلم
بان تنشف الجوهرة في عين الشمس فان رايت التغير شايعا
فيها كلها ظاهرا وباطنا ولا ضوء لها البته فلا تتعب

في علاجها فانها لا تنجلي وان رايت لها ضوءا فالوسخ طارى
عليها فتعالج بما اصفه ان كان اللولو صغيرا فيوضع في
لبن التين دايما يبدل له اللبن في كل ثلاثه ايام الى ان يتجلى
صفة اخرى يجعل في قدح مطين ويلقى عليه صابون وقلي
بالسوا ويوضع القدح على نار فحم لينه قدر ساعتين
او ثلث ساعات فانه يبين كاحسن ما كان صفة
اخرى يوخذ حليب مفشرجه ومثله من السمسم المقشور
ومثله من الكافور وتسحق الجميع وتعجنه قرصا وضع الحب
في وسطه واجمع القرص عليه وسده واجعله في مغرفه
حديد وصب عليه من دهن الاكارع [illegible] وغله بنار
خفيفه غليانا خفيفا فان الصفرة تخرج من الدر وتخرج
الحبه بيضا وان كان تغير اللولوء الى الحمرة فعله ساعة
في لبن حليب وخذ اشنانا فارسيا او شبا يمانيا وكافورا

8

التراب وقد جففت رطوبته فانه اصلح من ذلك ولذلك يخلطه
الكحالون في اكحالهم لنفعه وتشديده اعصاب العين
وخاصته مع ذلك النفع من خفقان القلب ومن الخوف
والجزع الذي يعرض من المرة السوداء ويلطف الدم الذي يغلظ
في الفؤاد ولهذا يخلطه المتطببون في ادوية القلب
ويحبس نزف الدم ويجلوا الاسنان طلاء صالحا واذا سحق
وسقي مع سمن بقر نفع من التسموم وذكر ارسطو
ان الحجر الذي يكون منه اللولو على ما قدمناه اذا قطر منه
في الكلف او غشي بعض اعضاء البدن البرص البس ذلك
العضو صبغا المداومة وذكر ايضا ان من وقف على
حجر الدر من كباره او صغاره حتى يصير ماء رجراجا
ثم طلي به البياض الذي يكون في البدن من البرص اذهبه
في اول طليه يطليه به وانه من كان به صداع من قبل

انتشار اعصاب العين وسعط بذلك الماء اذهبه وان
شفاه في اول تسعيطه قال احمد التيفاشي ثم
هذا الكتاب بما جربته واختبرته ووقفت عليه بالعمل
ان خاصة الاترج يحل الجوهر الا انه يحله حلا كالمني
لا يعلق بالاجسام اذا طلي عليها والمياه القوية الحريفة
تحله رجراجا تعلق بالاجساد ذكر قيمته وثمنه
العقد المتعارف عند اهل بغداد ستة وثلثون حبة
اقل العقود زنته سدس مثقال وهي اربعة قراريط قيمة
عشرة عقود من هذا العقد ثلثة ارباع دينار عقد
ربع دينار عقد ربع مثقال عشرة عقود بدينار عقد
ثلث مثقال عشرة عقود بدينار وربع عقد نصف مثقال
عشرة عقود بدينار عقد ثلثة ارباع مثقال العشرة عقود
باربعة دنانير عقد مثقال وربع خمسة عشر العشرة

9

عقد مثقال ونصف عشرون العشرة عقد مثقال ونصف
وربع خمسة وعشرون العشرة عقد مثقالين خمسة وثلثون
العشرة عقد مثقالين وربع باربعين دينارا العشرة
عقد مثقالين ونصف بخمسين العشرة عقد مثقالين
ونصف وربع وتسعين العشرة عقد ثلاثة مثاقيل ثمانين
العشرة عقد ثلثة مثاقيل وربع تسعين العشرة عقد
ثلاثة مثاقيل ونصف بماية وعشرة العشرة عقد ثلاثة
مثاقيل ونصف وربع بماية وخمسين العشرة عقد اربعة
مثاقيل بمائتي دينار العشرة فان كان نهاية في الجودة
والصفات التامة كانت قيمة العقد الذي زنته اربعة
مثاقيل كل عشرة عقود بثلثماية دينار لكل عقد ثلثون
دينارا ويخرج بعقوده ايضا حينئذ من باب العشرات
الى باب الاحاد فتكون قيمة العقد الواحد الذي زنته اربعة

مثاقيل ونصف اربعين دينارا او عقد اربعة ونصف وربع
بخمسة وخمسين دينارا او عقد خمسة مثاقيل بخمسة وستين
دينارا او عقد خمسة ونصف بخمسة وسبعين دينارا او عقد
خمسة ونصف بخمسة وثمانين دينارا او عقد خمسة ونصف
وربع بتسعين دينارا او عقد ستة بماية دينار وعقد سبعة
بماية وخمسين ويتضاعف بهذه النسبة الى انتهى ما يوجد
منه في الوزن والغبطة فيه بحسب جودته اوصافه الحسنة
وهي النقاوة والشفوف وتسمى عند الجوهريين الماسة وكبر
الجرم والدحرجة وضيق الثقب في الجواهر المفردة
افضل الجواهر المفردة الجوهرة التامة وهي المستديرة
الشكل في جميع جهاتها المستوية التي لا تضريس فيها ولا طول
ولا تسطيح ولا اعوجاج والجوهرة التي بهذه الصفة تسمى
عند عامة الناس المدحرجة وتسمى عند الجوهريين خاصة

6

الساحل استخرجوا ما فيه وذكر المسعودي انه اذا كان
شهر نيسان نزل على البحر الذي فيه صدف الدر مطر غزير
فيصعد ذلك الصدف ويفتح فاه لقطر الماء فاذا التقمه
غاص به على الوجه المذكور قيل وغاصة هذا البحر تكون
عندهم قوارير فيها دهن له في الماء بريق فاذا راوا حيوانا
موذيا ارسلوا منه شيا في البحر فانفرج له البحر صاعدا
فتراه تلك الحيوانات فتفزع منه وتنفر عنه وقيل وغاصة
هذا البحر ينبحون على تلك الحيوانات نبح الكلاب فتنفر من
ذلك مع بقية الاسباب التي يحتاطون بها المذكورة
فيما سلف وذكر يوحنا بن ماسويه في كتاب في
الاحجار ان الغايص لا يبعد في الغاصة حتى يخرق ما بين
اذنيه وحلقه فينبعث دما ثم يتمرن ويستمر فيكون تنفسه
منه تنفسا ضعيفا فاذا غاص جعل على نفسه ملزما عاج

او قرون لئلا يدخله الما وتنفس من موضع الشق ويضاربكت
الماء مقدار نصف ساعة ويغوص في اليوم ثلاث غطسات
من بكرة الى انتصاف النهار وغداوهم السمك والتمر
ذكر معدنه التي يتكون فيه الجوهر يوجد في معادن
كثيرة الا ان معظم الجيد منه الفاخر النفيس سرنديب
ثم كيس وعمان والبحرين وجزيرة خربته من كيس والبحرين
من ارض فارس وجوهرها افخر اصناف الجوهر وما يوجد
منه بغير هذه المواضع فلا اعتبار به وكذلك ما يوجد منه
ببحر القلزم وساير بحار الحجاز واليمن فردي ولو كانت
منه في بنادق الكبر فانها لا يكون لها طايل ثم اذ ليس
فيها شي من اوصاف الدر النفيس الذي ياتي ذكره بعد
وما يوجد منه في الاعماق والمواضع النقية من الكدر انقا
والون وسبب ذلك ان ما قرب من سطح الماء نفذت

7

اليه حرارة الشمس واثرت فيه صفرة وما كان في المواضع الحمئة
غيرته ابخرتها الرديه ولا تنجلي صفرته لانها اصلية من اول
تكونه شايعة فيه من باطنه الى ظاهره فاما ما انسخ منه
بالاستعمال فانه تنجلي صفرته وسنذكر ذلك فيما يستقبل
هذا الباب ذكر جيده ورديه ذكر جيده
ورديه الجوهر الكاملة خواصه اما في الكمية فالعظم
والوزن واما في الكيفية فشدة البياض وكثرة الماء
والاشراق واستواء اللون واستوا الاستدارة وشكله
واكتنازه وما لم يكن كذلك فالافات فشدته منها انه
ربما وجد بعض الدر لم يتم تربيتها وربما تعلق بها قشر من
لحم الحلزون وصار كالصدا والوسخ وانسد لونها وربما
كانت كدرة او كان فيها ماء او كانت فيها دودة او
كانت مجوفة غير مصمتة وكل هذه فهي افات دخلت

على الدرة في مقر التربية لها واما فساد شكلها من قبل ان
الحبة يقع في موضع من اللحم الذي في موضع الصدف غير
مستو فتتجسد الدرة على صورة الموضع الذي ضمها وجيد
الجوهر في الجملة هو المدحرج القاتا الصافي الشفاف الكبير
الجرم الكثير الوزن الضيق الثقب الجيد اللون الابيض
النقي من الوسخ ذكر خواصه ومنافعه
من خواص الجوهر في نفسه ان تكون قشورا ربما طبقة
على طبقة وما لم يكن كذلك فليس بجوهر مخلوق بل
مدلس مصنوع والجوهر بالجملة هو الدر وهو كبار
اللولو وحبه الصغير الذي لا يمكن ثقبه لصغره كل ذلك
فصول في الحر والبرد والرطوبة واليبس لطيف يجفف
الرطوبة في العين وينفع من ظلمة البصر والبياض في العين
وكثرة وسخها ولا سيما العتيق منه والذي يوجد في القبو

4

اذ وذ القوافي عني ذيادا ذياد آمر قد تحرا جرادا
فلما كثرن وعنينه تخير منهن شتى جيادا
فاعمل منها لها حاسه واحمد من دونها المستجادا

ان تكون في معدنه حيوان الجوهر الذي يسكون فيه كبيرة وصغيرة يسميان باليونانيه اسطوروس ويعلو الجم ذلك الحيوان صدفات ملتزقات بجسمه والذي يلي الصدفتين من جسمه اسود ولهذا الحيوان فم واذان وشحم من داخلها الى عامة الصدفتين والباقي رغوة وزبد وما ذكر ارسطو في كتابه في الحيوان غير الناطق ان السرطان يشتهي اكل لحم هذه الدابة فلما حال دونه ودون شهوته شيء بمنزلة السور الحاجز بينه وبين ذلك اللحم الرخص الذي في الصدف احتال عليه فلا يزال السرطان راصدا له حتى يراه قد فتح

حلد

جلد الصدفتين فياخذ حجرا صغيرا ويرمي به في جوف الصدفه فلا تطبق عند ذلك على انضمامها كما كانت فيدخل السرطان عند ذلك فريسه الى ذلك اللحم الرخص فيستخرجه وياكله لا لتلذذه باكله ويذكر من اكله من الغواصين انه شبيه الطعم بطعم قوانص الطير وذكر ارسطو في كتابه في الاحجار ان البحر المحيط بالعالم المسمى اوقيانوس في اوقات فصل الشتاء يهيج هيجانا شديدا فيطلبه الصدف الذي يكون فيه الدر وتهب ريح الشمال فاذا هبت الرياح والامواج من ذلك البحر المحيط دار لامواجه رشاش ووقع في البحر المسلوك فالتقمه الصدف الكاين في البحر الذي يسلكه الناس كما يلتقم الرحم النطفه من مني الرجل فتصير تلك النطفه من ذلك الما في اللحم المركب في الصدف فلا يزال الصدف يعلو الى الموضع الساكن

5

في البحر فيفتح فمه ويستقبل بذلك الماء الذي ينعقد مثل النطفه برياح الهواء وحر الشمس عند طلوعها وغروبها ولا يعرض لها في وسط النهار لشدة حر الشمس وهيجان البخارات التي تهيج من العالم والغبار الذي تهيجه الرياح فاذا انعقدت الدرة وصلبت وصارت جسدا استويا هبط الصدف الى قعر البحر فانغرس هناك في ارض البحر ويضرب بعروق وتشعب منه مثل الشجر ويصير نباتا بعد ان كان حيوانا ذا روح ونفس وفعل متحرك فيقطع مثل الثمرة المنضجة اذا انقطعت من الشجر وذكر ارسطوا ان الدرة ان تركت حتى يطول بها المكث تغيرت وضمرت وفسدت كالثمرة اذا ابقيت في الشجرة لم تقطف في وقتها ذهب نضارتها وطيب مطعمها فياتي الغايص الى خشبات المثقل مغروسه في موضع الدر يعلم بها الغاصه

الى المواضع

الى المواضع التي جرت العادة ان يكون الدر فيها فاذا راى الصدف اوقف مركبه قايما ويدلي حبلا من ليف لثقل ارضم فيه حجر يثقله ان كان للما حركة ثم يشد في الغايص حبلا وينزل مشدودا به حجر وزنه ستين رطلا او نحو ذلك بين عجان نسود ليفزع من سواد هذه الحيوانات المهلكة للغاصه فقد ذكر ان هذا البحر فيه حيوان ربما ابتلع الغايص وفيه حيوان يقال له القرش ربما قطع الغايص وفيه حيوان يقال وضروب من الحيوانات الضاره مانعا لها ولو جعلوا الحجر ابيض لظننته تلك الحيوانات طعما لها فاسرعت اليه فابتلعته وقطعت الحبل فلذلك يجعلون الحجر اسود فاذا غاصوا ووصلوا الى الصدف قطعه الغواصون بما هو مهيا لذلك مثل المناجل من اصله ورفعوه في مخالبهم كالشبكة من شريط ليسيل منها الما ويبقى الصدف فاذا اخرجوه الى

2

بسم الله الرحمن الرحيم
الحمد لله الملك الجبار العزيز القهار خالق
السموات والارض وما فيهما من عجايب الاثار وغرايب
الاسرار المودع معظمها في جواهر الاحجار الموجودة
في خزاين الملوك الكبار والروساء ذوي الاقدار
وصلواته على نبيه المصطفى المختار وعلى اله الابرار
وبعد فهذا كتاب غريب الوضع عجيب الجمع
عظيم النفع ضمنته ذكر جملة من جواهر الاحجار التي لا
يكاد يستغني عن اكتسابها ملك كبير ولا رييس خطير
لما تشتمل عليه من عجايب المنافع وعظايم الخواص وجميعها
متيسرة الوجود ولم اشرك بها ذكر شي من الاحجار الشاذة
الاسما النادرة الوقوع المتعذرة الوجود اذ كان ذلك
مما لا طايل تحته ولا جدوى في ذكره وانما نقتصر بذكر الحاصل في
الوجود

الوجود لا الباطل في حيز المعدوم المفقود وترجمته بازهار
الافكار في جواهر الاحجار وجملة عدة الاحجار
المثبتة فيه خمسة وعشرون حجرا وهي الجوهر الياقوت
الزمرد الزبرجد البلخش البنفش الحادي الماس
عين الهر البازهر الفيروزج العقيق الجزع المغناطيس
السنباذج الدهنج اللازورد المرجان السبج الجمست
الخماهان اليشم اليشب البلور الطلق وسبيلنا
ان نتكلم على كل واحد من هذه الاحجار المعدودة بعد
الالمام بشرح لغة ما له لغة منها في لسان العرب من
خمسة اوجه الاول علة تكونه في معدنه الثاني ذكر معدنه
التي يتكون فيه الثالث ذكر جيده ورديه وخالصه ومغشوشه
الرابع ذكر خواصه ومنافعه الخامس ذكر قيمته وثمنه على
اغلب الامور واوسط الاحوال ليكون هذا الكتاب بذلك

3

زايد مزية على الكتب الموضوعة في هذا الفن من عدة وجوه
اذ الكتب الموضوعة فيه اما ان يذكر فيها علة تكون الاحجار
ككتب المعادن واما ان يذكر فيها منافع الاحجار ككتب
الخواص واما ان يذكر الامرين معا ولا تعرض لذكر
قيمتها واثمانها لاجل ذلك كان هذا الكتاب اعم فايدة
واجدى عايدة من ساير الكتب الموضوعة في هذا الفن
فمعظم الخواص الموضوعة فيه مما جربته بنفسي او وثقت
بصحة النقل فيه عن غيري من المعتبرين واحلت عليه
منسوبا ذلك اليه وهذا اوان ابتدايه من الكتاب والله ولي
التوفيق الباب الاول في الجوهر
الجوهر لغة اسم عام لجميع الاحجار المعدنية النفيسة ثم
خص به هذا بعينه لفضله عليها كما ان الورد اسم عام لكل
زهر ثم خص به الورد المعروف لفضله عليها وكذلك

الغالية من اسما الجوهر الجماهان والسدر واحدته
جمانة وسدرة وهذه اسما لما ثقب منه فاما ما لم
يثقب منه فانه يقال له الدر والحب والخرايد واحدته درة
وحبة وخريدة وكذلك يقال للجارية البكر ما لم تفتض
خريدة ويقال ان اللولو اسم لما ثقب منه خاصة ويقال
للدرة والبيضة التومة ايضا والجمع توم مثل صورة وصور
وتومة وتوم مثل صورة وصور بالتحريك ويقال للدرة
الخريدة والجمع الخرايد والجوهر اسم عام يطلق على
الكبير منه والصغير فما كان منه كبيرا فهو الدر وسياتي
بيان ذلك وما كان منه صغيرا فهو اللولو المسمى في اصطلاح
الجوهريين حبا ويسمى ايضا اللولو الدق واللولو النظم
والمرجان في لغة العرب صغار الدر قال امرو القيس
ويقال انه اول شعر قاله

2773

Suppl. ar.
n: 880

English Translation with Explanatory Footnotes

Note from the translator

ARABIC is a very rich language. It has twenty-nine letters and a complicated grammar. The translation of some words may seem to be slightly different; this is because one word in Arabic can have several connotations, depending on the context and the sentence where it is used.

The Arabic used in *Azhar al Afkar fi Djawahir al Ahdjar* includes very few words of specialized vocabulary due to dialectical differences; for example, when al Tifaschi referred to the king of Tunisia, he called him king of Africa, as *Ifrikiyah* (Africa) used to refer to Tunis at that time; also, the word *Djawhar* meaning "pearl" is used only in North African countries, while in the rest of the Arab world, it usually means "jewel." But as the morphology and syntax of written Arabic is the same in all Arabic-speaking countries, a reader from, say, Yemen would have no problem reading a manuscript written by a Moroccan writer.

This is as much as possible a literal translation, in order to ensure faithful transmission of al Tifaschi's concepts and scientific knowledge of gemstones. Words in brackets are additional words used by the translator to clarify or complete the meaning when deemed necessary. In some instances, words or sentences from the Cairo version, arranged by Dr. M. Y. Hassan and Dr. M. B. Khafaga from the six different versions of the manuscript existing in Cairo, have been used to replace passages that are either missing or unclear in the Bibliothèque Nationale's version.

As to the style of the manuscript, what strikes the reader is the consistency and uniformity of the chapters. All chapter heads are numbered and bear as a title the name of the gemstone under study; they are done in thicker script than the rest of the book. Titles of paragraphs are also done in thicker script and are repeated under every chapter in the same order throughout. There is also consistency in establishing the prices of the various gemstones, although they were not done in tabulated form. Words such as "I say, I saw, he said, I was told, so and so said, they told me," have their last letter elongated to provide a clear reference to a quotation. Although the manuscript is a long one (eighty-five double pages), it was obviously written by one person, as it shows the same style of handwriting throughout.

The difficulty that faced me was two-fold: There were no punctuation marks, therefore the sentences needed careful reading in order to grasp the exact meaning; and there was almost no accentuation and little diacritical marks which made some of the unfamiliar words difficult to understand. However, on the whole, the manuscript was clear, and the scientific framework in which it was written simplified my job.

Samar Najm Abul Huda

BEST THOUGHTS ON THE BEST OF STONES

IN the name of Allah, the Merciful, the Compassionate. Praised be Allah the Mighty Ruler, the Esteemed, the Conqueror, Creator of Heaven and Earth and all the Wonders and Secrets therein, most of them lying in the precious stones found in the coffers of great kings and mighty rulers; Prayers be on His Chosen Messenger, al Mustafa,[1] and on His Worthy Descendants.

This book is a curious one, containing strange information, and is of great benefit. I have included therein a variety of precious stones that no great king or important nobleman can do without in view of their unusual benefits and great properties, all these stones being found in abundance. I have not mentioned any of the stones of eccentric name nor the rare and scarce ones, as it is of no use mentioning them. It is better to talk about what is available and not about what is lost; in other words, about Best Thoughts on the Best of Stones.

The stones described [in this book] are twenty-five in number, namely: Pearl, corundum, emerald, peridot, spinel, garnet, tourmaline, diamond, moonstone cat's eye, bezoar, turquoise, cornelian, agate, magnetite, emery, malachite, lapis lazuli, coral, jet, amethyst, hematite, jade, jasper, rock crystal, talc. We shall talk about each one of the above-mentioned stones, after explaining the origin of its Arabic name, according to [the following] five aspects: First, cause of formation of the stone in its host rock; second, characteristics of the mine [location] where the stone is formed; third, good and bad qualities, also genuine and imitation [fake] stones; fourth, properties and benefits of the stone; fifth, general average value and price.

This book will therefore provide more information than the [other] books written on this subject from many aspects, as the [other] books written on this subject either deal with the cause of formation of the stone, similar to those on minerals, or mention benefits of stones, such as books on properties of stones; or still they mention both these points but do not deal with their value or prices. For this reason this book is of greater benefit and better purpose than the other books written on this subject. I have tested myself most of the properties mentioned therein; or have ascertained those mentioned by other respectable authors, quoting the author in question [where appropriate].

This is where the book ends. May God grant me success [in my endeavor].

1. *Al Mustafa* is one of the many names of the Prophet Mohammad.

Chapter 1: On Pearl

THE term *djawhar* refers to all precious mineral stones, then it was also used to refer specifically to pearls as they are the most valuable of gems; similar to the word *ward* which is the general name for all flowers, but is also used to refer specifically to roses, as they are the best of flowers; likewise *al faghia.*[1] Of the many names of pearls are *Djumahan* and *Shazar,* plural of *Djumahana* and *Shazra,* these names referring to drilled pearls. As for pearls that are undrilled, they are called *Durr, Habb* and *Khara'ed,* plural of *Durra, Habba* and *Kharida.* Likewise, a slave girl still virgin, not yet deflowered, is called *Kharida.* It is also said that *Lou'lou'* is the specific name for drilled pearls. The white pearl is called *al Thuma,* singular of *Thuwam,* as in *sura* and *suwar,* and *Thuma* and *Thuwam* are like *sura* and *suwar* in accentuation.[2] Pearl is also called *Khafrada,* in plural *Khafared. Djawhar* is a general term that applies to both large and small pearls. Large pearls are called *Durr,* about which we shall talk later. Small ones are called *Lou'lou'* and in jewelers' terminology *Habb,* also *Loulou daq* and *Loulou nazhm.* In the Arabic language, *Mardjan* [3] refers to seed pearls.

Imru' al Kays[4] said, in what is supposed to be his first poem:

Poems come to my mind in waves
When they abound, I choose six good ones out of them,
Leaving the seed pearls aside . . . .

Cause of formation of the stone in its host rock

The pearl animal where large and small pearls are formed is called in Greek *astoros.* The meat of this animal is covered with two shells attached to its body, the meat sticking to the shell being of black color. The animal has a mouth and two ears and fat on its internal shell until the edges, the rest being foam and froth.

In his *Book on the Nonspeaking Animal,* Aristo[5] mentioned that the crab craves the meat of this animal, but as something stands like a barrier between this tender meat inside the shell and its craving, it reverts to a trick [to obtain it]. The crab keeps close watch, and when it sees the oyster opening its shell, the crab takes a small pebble and throws it inside, thus the oyster can no longer close as tightly as it could before and the crab thrusts its claws into the tender meat, removes it and eats it as it greatly enjoys this meal. Divers who have tasted it say that its meat is similar to that of birds' giblets.

In his *Book on Stones*, Aristo mentioned that the ocean called *Okananos*[6] that surrounds the world becomes strongly agitated in winter. The oysters that contain pearls seek this ocean during the times of the northern winds; and when the winds and the waves of this ocean become agitated, their spray reaches the navigated sea and is caught by the oysters that inhabit this sea, as the womb catches a droplet of the man's semen. Thus a tiny drop of this water goes into the meat within the oyster. The oyster then goes to a quiet place in the sea, opens its mouth, and receives this droplet of water formed from the wind and the heat of the sun during sunrise and sunset; it does not approach it during the midday hours, due to the strong heat of the sun and the agitation of the inhalations that are caused by the people[7] and the dust raised by the wind. If a pearl is born, solidifying and taking shape, the oyster goes down to the bottom of the sea and takes root in the seabed, where it grows byssae that ramify from it like trees, and it becomes a plant after having been an animal, with breath, soul, and movement; then it is picked like a mature fruit is picked from the tree.

Aristo mentioned that if a pearl is left too long in its shell, it changes, shrinks, and deteriorates like a fruit that stayed too long on the tree and was not picked in due time, thus losing its freshness and good taste. The diver approaches the wooden beams from the doom palm tree that were placed in locations known by divers to yield pearls; and when he sees the oysters, he holds his boat still and dangles a rope of fiber made from the doom tree or other, at the end of which is fastened a heavy stone for ballast in case the waters are agitated; then he dives along a strong rope to which is attached a stone of 60 *Ratls* or so, a black stone in order to frighten sea creatures harmful to the divers. He mentioned that, in this ocean, there is an animal that could swallow the diver and another one called *al qirsh*[8] that could cut a diver in two, as well as a variety of dangerous creatures. If the stone used was white, these animals would think it was some kind of bait and would rush to swallow it, thus cutting off the rope; for this reason the stone used is a black one. When the divers reach the oysters, they cut them from their roots with a tool specifically prepared for this purpose, similar to a sickle; and then put them in a sort of sieve made of material similar to a net so as to let the water out and keep the oysters in only. When they reach the shore, they remove the oysters from this net.

Al Mas'udi had mentioned that during the month of April, heavy rain falls on the sea where pearl oysters are found; these oysters rise to the surface [of the sea] and open their mouth, to catch drops of rain, after which they dive and return to the bottom of the sea in the aforementioned manner. The divers of this sea carry with them jars containing grease that shines in the water; and when they spot some dangerous creature, they throw some of this grease, which floats upwards in the water; and when they see it, these creatures get frightened and speed away. The divers of this sea emit sounds like dog barks at these creatures, which frighten them away, as do the previously-mentioned preventive measures.

Yuhanna ibn Masaweih[9] had written in his *Book on Stones* that the diver returns only when his eardrums explode and blood comes out. Then he keeps practicing until he is able to breathe faintly and when he dives, he puts on a nose clip made of ivory or horn so as to keep the water out, and he breathes from this opening [within his ears] and can stay up to half an hour underwater. He effects three dives daily from morning till midday, and his food consists of fish and dates.

On the mine [location] where it is formed

Pearls are found in many places, but the best and most precious ones are found in Sarandeeb[10] then Kish,'Oman, Bahrain, and the island of Kharq between Kish and Bahrain, in Persian lands. Pearls from these regions are the best, and what is found in other locations is of no value. Likewise what is found in the sea of Qulzum[11] and the seas of the holy Hidjaz[12] and Yemen is also of bad quality, and even if the pearl is a very large one, it is still of little value as it does not possess any of the qualities of precious pearls that will be mentioned later.

Pearls found in deep waters and in locations free of heat show better purity, sheen, and color. This is due to the fact that those that are closer to the surface of the sea are subjected to the heat of the sun, which gives them a yellowish tint. Pearls found in hot waters are affected by their negative vapors, and their yellowish shade cannot be removed as it is part of the pearl's initial constitution, coming from deep within it until its surface. Pearls that become dirty from usage can be cleaned and their yellowish tint removed, a process about which we shall talk later, after this section.

Good and bad qualitites

The perfect pearl is valuable either on account of its large size and heavy weight; or on account of its qualities: strong white color, good clarity and orient, even distribution and strength of color, completely spherical shape and perfect general appearance. Pearls that do not present these qualities possess inherent defects that have spoilt them; for example, lack of maturity, or maybe part of the snail's shell has stuck to the pearl and became like rust or dirt, thus ruining its color. The pearl can also be turbid or contain water or some worm, or can be hollow and not solid. All these are defects that have occurred during the formation of the pearl. As for the distorted shape,[13] it is due to the irregular place within the oyster meat where the pearl has formed, taking the shape of this location. The valuable pearl is the round, spherical, pure, translucent one, of large size and heavy weight, with a small drill hole, of good white color and free of any dirt.

Properties and benefits of pearls

One of the inherent characteristics of pearls is that they are formed of thin layers, one over the other;[14] and those that are not so are not genuine, but rather man-made, fake.

In general, *Djawhar* refers to *Durr*, including both large and small [seed] pearls, so small that they cannot be drilled. Pearls are average in heat, cold, moisture, and dryness; they are mild and can dry out moisture in the eyes, they are also beneficial for darkness,[15] white spots,[16] and dirt in the eyes; those pearls that are old and found in graves and in the sand, free of any moisture being the most effective. For this reason, makers of collyrium mix them in their collyrium preparations, as they are good for the eyes and strengthen the optic nerves. Pearls are also useful for heart palpitations, [they protect] against fear and anguish that result from the black bile (?), and reduce blood clots in the heart. For this purpose, physicians use pearls in their medicines for the heart. Pearls also can stop hemorrhage, give a good polish to the teeth, and if powdered and moistened with cow fat, are useful against poisons. Aristo had mentioned that if sea water from which pearls develop–as noted earlier–is sprinkled on the palm of the hand or if any part of the body is soaked in it, it acquires a coating similar to molten silver. He also mentioned that if a person dissolves large or small pearls until they become like agitated water, then spreads this solution over white leprosy spots, these disappear from the first application. Also if a person suffering from migraine due to spreading [or loosening?] of the optic nerves sniffs the above mentioned solution, he will be cured on the spot.

Ahmad al Tifaschi, the author of this book, said: What I have tried myself and tested empirically is that acid from the citrus medica can dissolve pearls although making them curdle like semen, not sticking if applied on bodies. Strong acrid water can dissolve them into an agitated solution that will stick to bodies.

Value and prices

In Baghdad, the regular necklace is made of thirty-six pearls, the lightest weighing 1/6 *Mithqal* or 4 carats, the value of ten such strings being 3/4 dinar.

A necklace of 1/4 *Mithqal,* 10 strings at 1 dinar
A necklace of 1/3 *Mithqal,* 10 strings at 1 1/4 dinars
A necklace of 1/2 *Mithqal,* 10 strings at 1 dinar[17]
A necklace of 3/4 *Mithqal,* 10 strings at 4 dinars
A necklace of 1 *Mithqal,* 10 strings at 10 dinars
A necklace of 1 1/4 *Mithqals,* 10 strings at 15 dinars
A necklace of 1 1/2 *Mithqals,* 10 strings at 20 dinars
A necklace of 1 3/4 *Mithqals,* 10 strings at 25 dinars
A necklace of 2 *Mithqals,* 10 strings at 35 dinars
A necklace of 2 1/4 *Mithqals,* 10 strings at 40 dinars
A necklace of 2 1/2 *Mithqals,* 10 strings at 50 dinars
A necklace of 2 3/4 *Mithqals,* 10 strings at 70 dinars
A necklace of 3 *Mithqals,* 10 strings at 80 dinars
A necklace of 3 1/4 *Mithqals,* 10 strings at 90 dinars
A necklace of 3 1/2 *Mithqals,* 10 strings at 110 dinars
A necklace of 3 3/4 *Mithqals,* 10 strings at 150 dinars
A necklace of 4 *Mithqals,* 10 strings at 200 dinars

If the necklaces are of exceptional quality and orient, the price of a necklace weighing 4 *Mithqals* is 300 dinars for ten strings, each string at 30 dinars.

Then the necklaces jump from the division of the dozens to that of the hundreds. Thus, the price of a single string weighing 4 1/2 *Mithqals* is 40 dinars, that of a string weighing 4 3/4 *Mithqals* is 55 dinars; that of a string weighing 5 *Mithqals* is 65 dinars;[18] that of a string weighing 5 1/4[19] *Mithqals* is 75 dinars;[20] that of a string weighing 5 1/2 *Mithqals* is 85 dinars; that of a string weighing 5 3/4 *Mithqals* is 90 dinars; that of a string weighing 6 *Mithqals* is 100 dinars; that of a string weighing 7 *Mithqals* is 150 dinars.

Prices double according to the above proportion until the heaviest and most sought after necklace is reached, depending on the quality of the five attributes, namely purity, translucency, called by jewelers "water," large size, sphericity, and smallness of the drill hole.

On the single pearl

The best single pearl is the "*Farra,*"[21] which is the round, spherical one, perfectly round from all sides without any blemishes or elongation or irregularity or distortion. The pearl that fits this description is commonly called spherical, and gem dealers refer to it specifically as the "running." It is pure in color with a good water, which is the sheen, of translucent core, which is the orient. Pearls that possess these qualities are called in jewelers' terminology *ratb.*

If a pearl possesses these qualities and weighs 1 *Mithqal,* it is worth 300 dinars. If two pearls weighing each 1 *Mithqal* possess these qualities and are so similar that one cannot differentiate between them, then their [combined] value is more than 700 dinars as they form an identical pair. If both pearls possessing the above qualities weigh jointly 1 *Mithqal,* their value is 100 dinars; if they weigh 2/3 *Mithqal,* their value is 27 dinars; if they weigh 1/2 *Mithqal,* their value is 20 dinars, and if they weigh 1/3 *Mithqal,* their value is 5 dinars.

Prices of pearls can be bargained according to demand, but inherent defects cannot be overlooked and cause not only some of the price, but most of it, to drop sharply. The reason for this is that the benefit derived from pearls comes from their beauty and appearance, not from any other property, as is the case with other stones. Pearls are wanted for adornment purposes; therefore, if beautiful appearance and shape are missing, there are no other reasons for enjoying them. You should know that any pearl over 2 *Dirhems* in weight, even if only by 1 *habba* [grain] per single pearl, is called *Durr* in jewelers' terms; and any pearl under 2 *Dirhems* in weight, even if only by 1 *habba,* is called *Djawhara* and *Habb* in their terminology. This is provided the pearl possesses all the right qualities. If its weight is more, but it shows any defect from the aforementioned ones, then it is also called *Habb* in their language, but its weight is not taken into consideration when it does not possess all the good qualities.

A pearl of 2 *Dirhems* and 1 *habba* or 2 in weight, for example, if it possesses all the previously-mentioned conditions of quality, is valued at 700 dinars. He said: If two pearls possess the above qualifications, they are valued at 2000 dinars each, provided they are a matching pair.

Defects in pearls

Scaling, irregular shape, yellow shade, off-white and unattractive white color, dull luster, large drill hole, small size, and light weight.

Substances that can harm pearls

All kinds of greases and acids, especially lemon juice, proximity to a fire, perspiration, bad odors, and contact with rough substances.

On polishing pearls

When polishing pearls, one should first take into consideration whether the change is present in its own constitution or whether it occurred after usage. This is known by examining the pearl under the light of the sun: If you see that the change is all over, inside and out, and it has no sheen whatsoever, then do not bother treating it, as it cannot be polished. But if you detect some light, then the dirt is only superficial, and you will treat it as I describe: If pearls are small, they should be soaked in milk from the fig tree, the milk to be changed every three days until the pearls are clean.

Another prescription: The pearls are placed in a clay container, equal amounts of soap and alkali are added, and the container is placed over a medium hot coal fire for two or three hours. The pearls will be better than they were before.

Another prescription: Equal portions of peeled *mahlab*,[22] peeled sesame seeds, and camphor are powdered together, then the dough is kneaded into a ball. Place the pearls in the middle, wrapped by the dough, which you will then seal [from all sides]. Place it in an iron ladle and cover with animal fat until it is fully immersed. Place on a low fire until it boils slightly. The yellow tint will come out, and the pearls will become white. If the pearls have a reddish tinge, you should boil them for one hour in milk, then take potash from Persia, or equal portions of alum from Yemen and camphor, powder them finely and knead them with some milk. Thickly coat the pearls with this mixture, then place them in some dough in a [clay] oven until the mixture is ready. The pearls will become white. But if they were spoilt by a coating of some sort, then take a portion of soap, an equal one of pure coarse salt of Andreen,[23] and another one of lime. Put them in a glass container, add clear water, and put the pearls in this mixture. Put over low coal fire, constantly removing the soap bubbles as they form along the way and renewing the water. Repeat the process several times, then wash the pearls with pure, limpid water. They will become clear and white and better than they were before.

Another prescription for polishing pearls: Powdered camphor is placed in a flimsy linen rag; the pearl is placed in the middle after pounding the camphor into a fine powder. The rag is tied and placed in a glass jar containing grease of *mahlab* seeds or of mercury, over a low coal fire for the time needed to count to 500. The pearls will come out clean, but if they still need more treatment, the process is repeated, and the pearls will be polished.[24]

Another prescription: Chalk is used instead of camphor, the process being the same, using grease of *mahlab* seeds and mercury.

Another prescription: Distilled acid from the citrus medica or strong distilled wine vinegar can clean pearls with a superficial yellowish tinge. This has been tried.

Another prescription: [Take] 1 carat of ammonia, 2 grains of tenkal, 1 grain of borax and 3 grains of alkali. Pound all the ingredients and place them in an iron ladle, then pour a sip of good red vinegar on the mixture. Boil slowly, then place the ladle in cold water without letting any water come into contact with the medicine. When cool, add a little powdered salt of Andreen, then take this medicine in the palm of your hand and rub the pearl with it thoroughly, then wash with clear water. The pearl will come out to your liking.

Notes

1. *Al faghia* is the sweet-smelling flower.
2. Meaning in the way these words are pronounced.
3. Nowadays, *Mardjan* refers only to coral.
4. Imru' al Kays is the most famous poet of the *Djahilliyya* (pre-Islamic) period.
5. Aristotle.
6. *Okayanos* in the Cairo version.
7. Or: the world.
8. Shark.
9. Eighth-century physician and translator of many medical books into Arabic.
10. Sri Lanka.
11. The Red Sea.
12. Saudi Arabia.
13. These pearls are now known as baroque pearls.
14. This is a reference to the concentric nature of pearls.

15. Myopeia?
16. Styes or albuginitis?
17. 2 dinars in the Cairo version.
18. 75 dinars in the Cairo version.
19. Literally: A necklace of 5 1/2 *Mithqals*.
20. 80 dinars in the Cairo version.
21. Meaning the running one.
22. *Prunus Mahaleb*, (Perfumed Cherry).
23. A village in Syria.
24. Meaning they will recover their original white color.

Chapter 2: On Corundum

ORIGIN of the name

It is also called *Djawhar, Kibrit,* and *'Asdjad* in some languages.

Cause of formation in its host rock

Pliny[1] mentioned in his book *Nature's Secret on Causes and Effects*[2] that stones come in many varieties and different colors; some are clear, others turbid; some are clear and hard, others are brittle; some melt in fire, others don't; some calcinate, others don't. They all originated from water and sand in larger or smaller proportions, depending on the place of their formation, also depending on how much or how little sunshine reached them in their host rock and how protected from the sun they were. This has resulted in their characteristics of strength, brittleness, different tastes, smells, and colors.

The cause of formation of corundum is that when the sun shone on the earth, it heated it with its strength, and the substances reached by the sun rays became heated, and the place [host rock] was strongly heated by the sun. The sun evaporated the moisture of the place where it shone; therefore when it became harder[3] due to little moisture, its power drew heat and hardness from the power of the sun, thus changing in nature, color, and taste according to the greater or smaller amount of moisture it contained previously. When the moisture was trapped over and around it, the water attracted what heat and dryness from the sun there was in this place, then the sun shone on it and heated it [few unclear words], reaching the dry areas. The water became hot and was thus able to dissolve the dryness that was accepted by the earth due to the heat of the sun, which led to its dissolution. Then the heat increased until it became extremely dry,[4] producing the stone called corundum, the extreme dryness of which drew its parts together; and due to its strong contraction, complete dissolution and strong moderation, it formed back into a solid, and its various parts were sintered together due to its dryness. This is the cause of the formation of corundum.

As to the variety of its colors, it depends on the different locations; if the water falls and remains in some place, it is affected by the substances that dissolve from the dry earth under the heat of the sun. The water is heated according to the amount of heat and solidifies as a red solid; or may be as a yellow one due to the little amount of heat present in it; or may be the heat was moderate in softening and dissolution, thus the substance solidified as a pure white solid. If drought conditions increase on the mine, cold prevails due to the overwhelming white color, and the presence of little heat, thus a black shade results and appears on its surface, reaching the red color within it, but the red could still project its color outside, despite the black shade on the surface; thus a blue color results. This is due to the yellow shade of the moisture that mingled with the black of the dryness, and that resulted in the blue color.

Pliny said that corundum is a gold base stone; and all stones except the base metals started by being corundum, as all the base metals started by being gold, but were prevented from becoming gold by different factors. Likewise, stones may have all started initially to be corundum, but were prevented from becoming so by excessive or little humidity and by excessive or little dryness. Instead, they became red or white or green or yellow or other colored stones that do not melt under the action of the fire and can be filed [scratched] by iron; while others are not, being called by many names other than corundum.

On the mine [location] where it is formed

Corundum is brought from a mine called Sahran[5] in an island behind the island of Sarandeeb at a distance of approximately 40 parasangs; the island itself being about 62 by 62 parasangs. There is a formidable mountain called Djabal al Rahun[6] from which winds and rains carry the corundum stones, which are picked up as pebbles from this place, also as alluvial deposits. It is said that if the sun shines on this mountain, it sparkles as the sun rays reflect off the corundum stones; this is called the lightning of the Rahun. This mountain is the one where Adam–may God bless him–descended from paradise and onto the earth. If this pebble is found,[7] it has a dark outer appearance with an overwhelming blackish dusty tinge similar to stones found in these colors in our country. If it is examined under the light of the sun, its red–or yellow or blue–or other color of the many colors of corundum will then be apparent.

Ahmad al Tifaschi–may God have mercy on his soul–said: Merchants who visited the island of Sarandeeb told me that the people of this place revert to the trick that we shall mention hereinafter in order to obtain these stones if winds and rains fail to bring them as usual during certain years. For this mountain where the corundums are found is a formidable one, difficult to reach and inaccessible on the top. Many eagles live and build nests on top of this mountain, as it is uninhabited by mankind; therefore, the people of this place slaughter an animal, skin it, and cut it into large pieces, then leave it at the foot of this mountain; then they go far and keep a close watch. Eagles come to retrieve this meat and carry it to their nests; and if they place it on the ground, the corundum stones stick to it. Then other eagles gather around the meat trying to steal it; some carry it away from the mountain, and the corundum stones fall off due to their heavy weight, to be picked up by the people watching the place where they have fallen. It is told that there are wild jungles, deep trenches, and towering trees at the foot of this mountain, inhabited by huge snakes that could swallow a human being or the head of a cow or other animal whole. If they do, they go to the trunk of a tree, wrap themselves tightly around it in order to break what they swallowed to pieces in their stomachs, thus digesting it. For this reason also, it is impossible to ascend this mountain or reach the strange stones found there. Some corundums are found in this island in the gravel and mud, but only the best stones carried by the rain from the mountain are taken, the sign of good quality being great luster, color, and brilliance. Rubies come from all these locations, some showing a delicate red color, others a deep red one. The deep red ones show a black tinge, and the red color appears only as veins within this black [background]. The stone can also have some empty cavity containing air, or even also water or mud. All this can be seen from the exterior. Therefore, the drill is used on it until this defect is removed, and the water, air, and mud come out. The stone could also contain holes similar to wormwood[8] called *Rim*,[9] the drill is then placed on the side of the inclusion until it is removed. If any defects are left within the stone, it could split.

Most [rubies] are found with an overwhelming blackish tinge, which comes out with heat treatment. Their method is the following: Stones from the mother rock where ruby is found are taken, powdered, dissolved in water, then pounded[10] until they are sintered together. They are used to cover the stone [to be treated] completely, then they are placed on a rock, other rocks are gathered around it, and dry wood is added on top. The fire is lit and is kept alive by constant blowing [with bellows] and adding of more logs of wood, according to the amount of black [in the stone], which they know from careful observation and experience. They blow on it from one single hour to twenty days and nights nonstop with the constant addition of wood according to the amount of black in the stone. If at the end of the treatment, the black tint is removed, and the stone has become any color other than black, they do not heat it again, because after the black shade is removed, the [residual] color does not increase or decrease. This is how stones are treated in Sarandeeb; some are also treated in Yemen, but this is rare.

Yuhanna Ibn Masaweih[11] said that ruby can often be found in sizes of 5 *Mithqals*.[12] He said that some sapphires may contain some yellow shade, which is removed by heat treatment.[13] But if it is overtreated, the blue color turns to white. If heat treated, blue and yellow sapphires turn to white and the white color [obtained] is irreversible. He said a sapphire can reach 40 *Mithqals*[14] in weight.

Varieties of corundums and description of good and bad qualities

Corundum comes in four varieties: red, yellow, blue, and white. Red corundum [ruby] is divided into four groups: The *wardi* [rose-colored] ruby, of a color similar to that of a pink rose, ranging in color and intensity from pink to white. Then the *khayri* [15] of a red color with a purplish shade, similar to the color of the cheirantus [gillyflower] or of stronger purple, varying in color and intensity until it is almost white. Then the red one of a color similar to the strong red safflower, also ranging in color and intensity from pink to strong, bright, safflower red. Then the *bahramani*, of pure red color, without any other shade, ranging in color and intensity from the color of the *wars* (?) until it reaches the color of the bright red safflower. *Bahraman* is one of the names of safflower after which this variety of ruby is called.

The most valuable ruby is the red one, which is the *bahramani.* The most valuable of all the other corundum varieties is the most intensely colored, the most delicate and transparent, with the brightest luster, free from the defects that we shall mention later.

As for yellow sapphire, we have a delicate one, light-yellow in color, very lustrous and bright; and the *khaluqi* [the color of turmeric], which is of stronger yellow than the delicate one. Also the *gulnari* [color of the pomegranate blossom], of stronger yellow than the *khaluqi;* it is the brightest, the clearest, and the best of the yellow sapphires.

Then blue sapphires; including blue, sky-blue [the color of lapis lazuli]; ultramarine, and navy-blue, which is of stronger blue than the ultramarine and is called *zayti*.[16]

The white variety includes the *maha'i*,[17] with the clearest and most intense bright white color and is the hardest. It is also the lowest variety and is the least valuable of all the corundums.[18]

On the purest corundum

The most valuable of the corundums are the red *bahramani,* the *rummani* and the *wardi,* of bright lustrous color, which are so completely transparent that one can easily see through them, free of any defect.

Defects in corundum

Crack lines and wormwoodlike inclusions. Crack lines are pseudo-cleavage lines that can be seen in the stone, and wormwoodlike inclusions are holes inside it, covered by some metallic dust. These holes could contain live, moving worms, the worm dying on contact with the air. This is what we were told by authorities on the subject.

Defects in colors

The worst color in ruby is the pink one with a white shade, and the *summaqi* [porphyry] one. The worst color in sapphire is the one with a greyish shade, called *sannuri;* also the one called *zayti*. The worst color in yellow sapphires is the one with a whitish shade. The worst corundums of all are the ones with ugly appearance, cracks, twinning,[19] and holes called *souss.*[20]

Nature of corundum and its power

The power of corundum depends on the mines where it is formed and on its color and varieties. Rubies are hot, dry stones; yellow sapphires are the closest to the red ones; they contain some heat, and for this reason they became yellow. Blue sapphires are whiter and more dry, and white sapphires are the coldest and most humid.

Characteristics of corundum

One property of corundum is that it cuts all stones similar to diamond,[21] but it can only be cut by diamond; for this reason a piece [point] of diamond is stuck on an iron drill, then it [the corundum stone] can be drilled in the same manner that wood is drilled.[22] Another property of corundum is that it cannot be polished on wood from the agigantea tree, which is used for polishing all stones except corundum. Corundum can only be polished on a brass lap. Agate stones from Yemen are broken to pieces, then burnt until they become like lime. Then they are moistened with water until they become like glue. The corundum stone is polished with this mixture on a brass lap. When polished, it has the highest luster of all stones. Another property of corundum is its brilliance, which no other transparent stone can attain. Another property is its heavy weight; it is the heaviest of stones similar to it in value. Another property of corundum is its resistance to fire; when submitted to high temperatures, it does not calcinate like other precious stones, such as emerald or others. Another property is that it can withstand a quick drop in temperature when removed from the fire [without being affected], different from most other stones. From all the different colors of corundum, only the red one remains stable when heat-treated, all the other colors, such as yellow, blue, and blackish are removed by the action of the fire, the stone becoming white and calcinating and cracking under excessive temperatures. The yellow color is the most resistant; the black one is the least stable under the action of the fire. The red color is not only the only stable one, but it also becomes more intense.

Aristo had mentioned in his *Book on Stones* that if one blows on ruby in the fire, it improves in appearance, and its red color is intensified; and if the ruby shows a deep red spot, this spot will spread out [evenly] into the whole stone upon heat treatment, thus improving its appearance; and if it shows a blackish spot, the black shade will be removed. The stone increases in beauty and clarity with heat treatment, and if a red stone loses its color upon heat treatment, then it is not a ruby but one of its imitations, and it is man-made and fake.

I have seen in the jewelry market in Cairo–may God protect her–stones that are being sold as genuine blue and yellow sapphires, while in fact they were manufactured and cheated, being originally white sapphires. Another property is that files and iron have no effect on corundum; nothing can scratch any of its varieties, whether red, yellow, or blue. Another property is that it can cut all transparent stones except diamond, no other stone being superior to any variety of corundum in all these properties.

Properties and benefits of corundum

One of the properties of corundum is what Aristo had mentioned; namely, that if any of the varieties mentioned above is worn in a necklace or in a ring, and this person is in a country hit by the plague, [he will not be affected like the people of this country].[23] Corundum also makes him nobler in the eyes of the people, facilitates his errands, and helps him overcome difficult problems in life. Corundum also strengthens the heart of its wearer and instills courage in him; it also makes him more esteemed and more respected in the heart of the people. Another property is that it is beneficial for heart palpitations and obsessive anxieties if it is hung. Another property is that lightning never strikes whoever wears it in a ring or in an amulet. Another property is that it prevents blood clots and stops hemorrhage, when hung. Another property, described to me by an honest jeweler well known for his experience and intelligence in this art who had gone to India and indulged a lot in the study of gemstones, is that people in India say that whoever possesses a corundum stone can attract to him powerful beings if these are not carrying such a stone, on the condition that this is done unintentionally and not for experimentation or testing purposes. One of the properties of yellow corundum is that it prevents erotic dreams.

There are stones similar to genuine corundum of nearly identical colors, as the above-mentioned corundum stones, though not reaching the brilliance nor having the properties of the genuine stones. The weakness of all imitations of corundums is that they can be scratched by ruby while it is itself not scratched by any of these stones. Also none of the imitations can resist heat treatment without burning. Ruby not only keeps its beauty and color when heated, but it also improves in appearance, as we have already mentioned.

Varieties of corundum and prices

You should know that the value and price of all stones depend on two factors: one inherent in the stone itself, the other due to external factors. As for the inherent factors, they are two, one is the good or bad quality of the stone, the other its large or small size. As for the external factors, they are supply and demand of the market and distance of the mines from the market. As for us, we are establishing the value of the stones mentioned in this book, according to their price in two marketplaces, namely, Baghdad and Egypt,[24] as they represent the average to which all others refer.

The price of a good, pure, flawless stone depends only on its large or small size, the price doubling according to the increase in size and decreasing with its decrease, in accordance with what I mentioned before.

The average pricc of a genuine *bahramani* ruby in Baghdad, only slightly superior in Cairo, is the following:

The price of a stone weighing 1/2 *Dirhem* is 6 pure gold *Mithqals*, each carat at 10 pure silver *Naqara Dirhems*, equivalent to 3/4 pure gold *Mithqals.*

The price of a stone weighing 1 *Dirhem* is 16 dinars, each carat at 2 dinars.

The price of a stone weighing 1 *Mithqal* is 2 1/2 dinars per carat.
The price of a stone weighing 1 1/3 *Mithqals* is 3 dinars per carat.
The price of a stone weighing 2 *Mithqals* varies from 4 dinars per carat to 3 1/2, depending on its color and increase in its clarity, the price rising according to the increase in its color, clarity, and large or small size.

The price of a stone weighing 1 *Mithqal* could reach 100 gold *Mithqals*[25] if the stone is a *bahramani* [ruby] of top quality, color, clarity, and brilliance, [having been] worked,[26] thus having lost a lot of its original size after working and polishing.

As for blue and golden[27] sapphires, their value is the same, each *Dirhem* at 4 dinars. As for yellow sapphire, each *Dirhem* is priced at 2 dinars. As for white sapphire, its value is half that of the yellow one. All this depends on increase or decrease in color and clarity, as we have mentioned previously; this difference due to the original place of formation, which we have described before.

I have read in some history books that the reason of Khalid ibn 'Abdallah al Qasri's[28] bankruptcy at the hands of Hisham ibn 'Abdel Malik[29] is because he [Hisham] was told that he [Khalid] had bought a ruby that could fit into one's palm for the price of 60,000 dinars. And when he [Hisham] asked for the stone, the latter refused [to give it to him], thus he [Hisham] caused his ruin and recovered the stone by confiscating his property.

Abu 'Abdallah al Ghalabi[30] mentioned in his book on *The Dignitaries of Quraish* when he wrote about Hisham ibn Abdel Malik that Yusuf ibn 'Omar[31] sent to the above-mentioned Hisham a ruby that could hardly fit in his palm, and the most magnificent pearl there could be. This ruby belonged to Rabi'a, the slave girl of Khalid ibn 'Abdallah, from whom he had bought it for 73,000 dinars.

Ibn Bakhtishew'[32] mentioned that Djaa'far al Barmaki's mother[33] gave him a bowl she used to eat from, together with its spoon, both of yellow sapphire, which he sold; and the price he sold them for was the cause of his lifelong richness.

Notes

1. Al Tifaschi called Pliny "Balinus" or "Blinus."
2. Pliny's *Natural History*.
3. Literally: drier.
4. Dry in the sense of hard.

5. Sahbaran or Sahayran?
6. Djabal al Rahun is none other than Adam's Peak in Sri Lanka.
7. Or: formed?
8. Al Tifaschi was probably referring to rutile needles, a common inclusion in rubies from Sri Lanka.
9. "*Rim*" or "*raym.*" Correct word is "*ratm,*" which means "impurities."
10. Heated instead of pounded in the Cairo version.
11. Eigth-century physician and translator of many medical books into Arabic.
12. 120 carats.
13. Literally: by placing it in the coal.
14. 960 carats.
15. The *khayri* flower is the cheiranthus, or gillyflower.
16. *Zayti* means here a navy-blue color.
17. *Al maha'i* comes from *al maha*, which is a stone of strong white color, similar to rock crystal.
18. In the Cairo version, another white variety called "the male" is also mentioned.
19. Al Tifaschi used the word *tara'eq*, meaning "lines" or "roads," to describe twinning, a common feature in ruby.
20. By *souss*, al Tifaschi probably meant rutile needles, that look like wormwood inside the stone.
21. This sentence could also be interpreted as follows: "It can cut all stones like diamond can."
22. Meaning as easily as wood can be drilled. This sentence can also be interpreted as follows: "then it can drill [other stones] as [easily] as wood is drilled."
23. The sentence between brackets is missing here. It was taken from the Cairo version to complete the meaning.
24. Meaning Cairo, as Misr (Egypt) is used locally to mean Cairo.
25. Literally: The price of a stone weighing 1 *Mithqal* of gold. Correct sentence is as mentioned above, which is taken from the Cairo version.
26. Meaning cut into some kind of cut, then polished.
27. Dark-blue in the Cairo version.
28. Khalid ibn 'Abdallah al Qasri, governor for Hisham in Iraq.
29. Hisham ibn 'Abdel Malik, A.D. 724-743, one of the Umayyad caliphs.
30. Unidentified person.
31. Unidentified person.
32. Ninth-century court physician of al Rashid, al Ma'mun, and the Barmakids.

33. The word “mother” is taken from the Cairo version. In this version, the name “misdaqa” was mentioned.

Chapter 3: On Emerald

IT is called *zumurrudh* with a *damma*[1] on the r and m letters, the r being stressed;[2] with a foreign dh[3] at the end, as the Arabs referred to it.

Al Farabi[4] said in his *Book on the Language*, that *zarbardjad* is the Arabic for *zumurrud.* This is an error, as *zabardjad* is another variety of stone about which we shall talk in the following chapter, God willing.

CAUSE of formation of the stone in its host rock

Pliny had mentioned that the emerald is the corundum and that it began forming as a ruby in all its parts and that it was red in color, but due to the strong density of the red color, it turned to black, thus becoming blue due to its dryness. The cause of the blue color is that the clear red color rose to the surface and became yellow, while the color remaining within stayed blue. As the heat increased, the stone was formed,[5] the two colors intermingled, and the outer and inner shades were mixed; giving birth to a green color; the stone thus became green and was called emerald.[6] It started by being a corundum because corundum is a gold base stone and is the origin of all stones, as gold is the leader of all molten substances. Also, it is like corundum, which cannot be melted by the action of the fire or scratched by iron, due to its dryness reaching until its surface resulting from the compact sintering of its parts; its dryness appeared at the surface, thus it resisted the file and iron; it became light in weight because the dry parts in it dissolved under mild and gentle cooking conditions. And when its dryness dissolved within it, it could not become compact again like corundum, thus increasing in weight and becoming hard; rather it took a long time to form under mild heat and gentle cooking conditions. When it [emerald] was completely formed, heat and drought affected it; it became compact and resisted the melting effect of the fire. This is the cause of formation of emerald and peridot in their host rock.

Pliny said: The corundum could be emerald, and the emerald corundum, as silver could be gold and brass silver, turning one into the other; as they started out from a common origin, but were affected by factors and thus changed in nature. He said: As we have described with the molten substances that change color until they revert to their original essence, likewise stones undergo the same changes as the substances.

On the mine [location] where it is formed

Emerald is brought from the borders between Egypt and Sudan, behind Aswan, from a mountain[7] that extends like a bridge from where minerals are extracted; thus small pieces of emerald are gathered, small stones similar to pebbles, deeply embedded in the host rock. The chief mineralogist in Egypt appointed by the sultan to be in charge of this mineral told me that the first thing that appears in the emerald mine is something called *talq* [talc], which are black stones that yield golden marcasite when heat-treated. He said: Then we continue digging, and we find friable talc where emeralds are found, embedded in a soft red earth. We could hit the vein, thus taking the emerald crystal whole, which is the best quality. As for the small stones, they are found by sieving the sand; the sand is washed [panned] as the sand where silver is found is washed [panned]; and stone after stone is found, some covered with a blackish layer similar to kohl; but they are small like mustard seeds, or slightly bigger; such are the ones this mineralogist expert showed me. Those stones found in the sand are called *fass*, and those cut off from the vein are called *qasab* in jewelers' and mineralogists' terminology, the latter ones being the oldest, the best, and the purest.

Good and bad qualities

Emerald comes in four varieties: *zhubabi, rihani, silqi* and *sabuni.* The most valuable, the most expensive, and the one with the best properties of emeralds is the *zhubabi,* of a very deep green without any other shade of color, of a beautiful green with good clarity. It is said that it was called *zhubabi* as its color is similar to that of the big spring flies [cantharides] that are found in the fields, not the small ones found inside the houses. They have the best green sheen, and this color is not found in house flies. As for the remaining above-mentioned varieties of emerald other than *zhubabi,* they are inferior in all aspects to the *zhubabi* one. The *rihani* has a light color like that of basil leaves, followed by the *silqi* which is of a chard green color, followed by the *sabuni,* of soap green color and of no value worth mentioning.

The best kind is the dark one with a whitish tinge, called *al 'Arabi* [the Arabian]; it is found in Barriyet al 'Arab[8] in Hidjaz;[9] and we have already mentioned that the best and most valuable emerald is the *zhubabi* one with the greatest clarity, without any tinge of yellow or black or other color; of great brilliance. If, in addition to the above, we have a big size and a straight undistorted crystal, then we have the top quality stone of greatest value.

Defects in emerald

The biggest defect in a *zhubabi* emerald is pleochroism, where one spot shows a different color from the other. Another defect is distortion of the crystal, this applying also to corundum and any other transparent stone, whether precious or not. Another defect is cracks [within the stone] of which emerald is virtually never free, being hidden[10] fissures within the stone.

Characteristics of the zhubabi emerald

The characteristic of the *zhubabi* emerald that is exclusive to it amongst all other stones and could be used to test a genuine from a fake stone is that the eyes of snakes pop out of their sockets on the spot when they look at it.[11] Ahmad al Tifaschi said: I used to ponder about this characteristic of emerald mentioned in the *Book on Stones*, then I tested it myself and found it to be true. For I had acquired a pure *zhubabi* emerald that I wanted to test using the eyes of snakes. Therefore, I hired a man to catch a snake, which he did, and I put it in a large basin. I took a piece of wax, stuck it at the end of an arrow, and glued the emerald on it, then I approached it from the eyes of the snake, which started leaping first at the arrow, moving furiously, trying to get out of the basin. As the emerald neared its eyes, I heard a light noise, as if one was killing a nit on his nail, and I saw that the eyes of the snake had bulged considerably, and it roamed perplexed in the basin, not knowing which direction to turn to nor wanting to get out. Most of its movements had become still, and its strength had abandoned it completely.

A characteristic of emerald is that it is soft and brittle, it is also light in weight due to its softness and brittleness. Another characteristic is that its green color increases and becomes clearer when it is foiled.[12] Another characteristic is its extreme smoothness and clarity and softness. The most specific characteristic of emerald is lightness of weight and inclusion of cracks, of which it is virtually never free, as we have already mentioned. Its light weight is due to its original formation, the cause of which we have mentioned earlier. Emerald disintegrates under the action of the fire and calcinates; it is not stable under high pressures like ruby, due to the above-mentioned reasons of softness and brittleness.

Properties and benefits of emerald

One of the properties is that it strengthens the eyesight if one stares at it for a long time; also, an emerald worn in a necklace or ring protects its wearer from the disease called epilepsy if it is worn before the disease occurs. For this reason, the sages used to order kings to make their children wear this stone around their neck at birth, in order to prevent the occurrence of epilepsy. Another property is that if an emerald of a weight equivalent to eight grains of barley is dissolved [in some liquid] and drunk by a person who has been poisoned before the poison starts taking effect, he will be saved from death, his hair and skin will not be torn out, and he will survive. Another property of emerald is that it is beneficial against blood discharges and against the affliction called dysentery if it is hung over the liver and the stomach externally. I have myself seen someone doing this in the *Maghreb,*[13] and he was cured; and I have prescribed this cure in the *Mashreq*[14] for sufferers from dysentery who had been unsuccessfully treated by doctors and were about to die, and when they wore this stone, they were cured, God willing. Another property is that it is beneficial for stomach pains if it is placed over the stomach externally. Another property is that it repels all poisonous animals, which do not approach the wearer of an emerald. Another property is that all varieties of emerald can be hung on the upper arm for protection and for talismanic purposes and on the thigh to speed childbirth. All this has been tried. Aristo said that emerald is colder and harder[15] than pearls, as this stone is taken out of the earth.[16]

A stone that resembles emerald is something called *al mash,*[17] which is found in the same host rock. It also resembles emerald in all its external characteristics, namely color, clarity, brittleness, and lightness of weight, and only gem experts can differentiate between the two stones. Its characteristic that differentiates it from emerald is that it loses its clarity and turns to blackish or yellowish if it is foiled, at which time it can be differentiated from emerald, which, as we have already mentioned, improves in beauty and clarity when any of its varieties is foiled. Other stones that resemble emerald are jasper, green jade,[18] peridot, and green sapphire.

Value and prices

You should know that all the properties and benefits described above are exclusive to the *zhubabi* kind of emerald, the properties of which make its price higher. The price of a pure *zhubabi* emerald weighing 1 *Dirhem* is 4 dinars per carat, the price doubling according to the increase in size, and decreasing according to its decrease, also in accordance with the previously-mentioned conditions of quality. However, the decrease in its price is less than with that of other stones, as it is a noble stone of great benefits, the properties being inherent in large or small stones, straight or distorted.

As for the remaining three varieties of emerald other than *zhubabi,* they are of no particular value, as they do not possess the properties of the *zhubabi* kind.

The wise judge Mu'in al Din Mubasshir, the superintendent of the emerald mines in Egyptian territories told me [the following]: I found in the emerald mine called Wadi al Shah, which is the biggest emerald mine between Qus[19] and Aizab, an emerald crystal of the *silqi* variety that had broken under the inadvertent blows of the mine worker while still embedded in the talc. I gathered its pieces and weighed them; they weighed 88 *Dirhems.* I found later in the same place, below where the crystal had broken, an emerald crystal of the *zhubabi* kind weighing 6 *Dirhems,* which I took to the royal Salihid coffer, may God raise the standing of its king. He said: I bought from a merchant in Qus, during the reign of the great sultan al Kamil[20]–God bless his soul–an emerald crystal of the *rihani* variety that he [the merchant] had found in a mine near his hometown. After working, grinding and polishing the stone, it weighed 12 *Mithqals.* He said: I had bought it from the merchant and taken it to the sultan al Kamil–may God bless his soul–in Damascus, where it was valued at 30,000 *Naqara Dirhems.*

Emerald mines are located at a walking distance of ten days from Qus, where there are a lot of active mines being worked until now; such as Qarshandah, Makamin,[21] Qu' Ghazi[22] and Wadi al Nahr,[23] which is a big mine. Wadi al Shah was so called because a copper statue of the shah was found there. The above-mentioned sheikh told me that humid formations resembling rust are found in these mines, and he showed me some black translucent stones he had obtained from these mines that could easily be polished, which he said were black emeralds; and other stones ranging in color between whitish and yellowish, brittle like emeralds.[24] The above-mentioned person had also discovered in Egypt, in a location at a distance of few days from Misr[25] itself called Tura,[26] ruby mines from which he showed me some red stones, small like mustard seeds or slightly larger.

Notes

1. *Damma* is a vowel sign placed over the consonants to render the sound "u."

2. To stress a letter in Arabic, a vowel sign called *shadda* is placed over it. This is similar to a double letter (ex. mm or dd) in English.

3. To give the sound "dh" in Arabic, a diacritical sign is placed over the "da" (d).

4. Wrongly written al Fasi in this version. Should be al Farabi (Alpharabius), a philosopher, physician, mathematician, occult scientist, and musician, who lived in the tenth century A.D., in the court of Sayf al Dawla al Hamadani.

5. Literally: cooked.

6. The word used was *zabardjad* (peridot), but this is obviously an error and should be *zumurrud* (emerald) for the meaning to be correct.

7. The word "mountain" is missing from this version. It was taken from the Cairo one.

8. Arabian desert.

9. Hidjaz, Saudi Arabia.

10. Meaning "internal" in this context.

11. As a matter of fact, Bailak al Qabadjaqi wrote in 1282 a lapidary entitled "The Treasure of Merchants with regard to the Knowledge of Gemstones, where he mentions that Djabir ibn Hayyan (eighth century A.D.), known in the West as Geber, the "father of Arab alchemy," mentioned this property of emeralds in his writings.

12. Literally: placed over a lining.

13. The *Maghreb* can mean either Morocco or all the North African countries.

14. The *Mashreq* covers all the countries from Egypt to Syria, including Lebanon, Jordan, and Palestine.

15. Literally: more dry than.

16. As emerald is a mineral and not an organic substance like pearls.

17. Unidentified stone that could be green beryl.

18. Nephrite.

19. Qus was at a time the seat of the Caliphate.

20. Sultan al Kamil, Ayyubid sultan of Egypt, A.D. 1218-1238.

21. Takayu in the Cairo version.

22. Fara'a Ghazi in the Cairo version.

23. Wadi al Bir in the Cairo version.

24. Literally: brittle as if they were starting to be emeralds.
25. Misr (Egypt) is used here to mean Cairo, a common practice in Egyptian popular use.
26. Tura, a village in the Gizeh district of Cairo.

Chapter 4: On Peridot

CAUSE of formation of the stone in its host rock

The cause of formation of peridot in its host rock is the same as that mentioned for emerald, peridot having started to be emerald, but due to factors that affected it such as lack of maturity[1] and lack of heat, it became softer and inferior in color, and peridot was formed.

On the mine [location] where it is formed

Peridot is found in the same mine as emerald, together with it,[2] but it is rare, much rarer than emerald. At the time when this book was written, which is the year 640 [A.H.],[3] no [peridot] stones were found in any mine, the few that are in the hands of people being small stones recovered while digging old archaeological sites in Alexandria–may God protect her–being the remnants of Alexander's treasures. The gem dealer and the claim owner who dug out these stones in Alexandria told me that he had recovered them from the above-mentioned place, and he showed me some of these stones. He said: I used to find the stone covered with a purplish layer that hid its true color, and when polished, it became extremely clear and bright. I[4] had seen with this person a stone of approximately 1 *Dirhem* that one could not remove his eyes from, on account of its delicate clarity, beautiful color, and purity, which he said he had recovered by digging at the above-mentioned sites in Alexandria.

Good and bad qualities

Peridot comes in dark green, light green, and medium green, of good clarity, delicate and transparent, which can be traversed quickly by eyesight.[5] This is the best kind.

Properties and benefits

Peridot does not possess any of the aforementioned properties or benefits of emeralds; it does not possess any characteristics except that it is a beautiful transparent green-colored stone, and it improves and strengthens eyesight when one stares at it.[6]

Value and price

1/2 *Dirhem* of pure [genuine] peridot is worth 1 dinar.

Notes

1. Literally: lack of cooking time.
2. "Is found in it" in this version, but the correct sentence should be "together with it," as it is written in the Cairo version.
3. This confirms the date at which the book was written, namely A.D. 1240.
4. Meaning al Tifaschi here.
5. Meaning completely transparent.
6. In the margin next to this text, is written the following note: "Not in original (?). It is said that if powdered peridot is drunk, it is useful against leprosy. This is a great benefit that the author [of this book] has neglected; may God have mercy on his soul.

Chapter 5: On Spinel

CAUSE of formation of the stone in its host rock

The author of this book said: Spinel and garnet and tourmaline are three stones similar to corundum, as peridot and *mash* are similar to emerald. The mode of formation of the above-mentioned three stones is similar to ruby,[1] and they are found in locations near one another.

Pliny mentioned on the cause of their formation that red or wine-colored stones like cornelian and tourmaline started by all being corundum, but increase or decrease in moisture and dryness as well as lack of moderation affected them and prevented them from becoming corundum; instead, they became red and wine-colored stones that are resistant to high temperatures like corundum; but can be scratched by corundum. They were given different names as they are different stones, all these stones having started to be corundum but due to external factors they did not continue, and their comparison to corundum is like that of corundum [to gold, which is the leader of all the minerals].[2]

On the mine [location] where it is formed

It comes from Balakhshan, foreigners say Balakhshazh with a "zhal," which is one of their towns [unclear two words][3] on the border with China, where the mines of this stone are in a large province. A Persian merchant who visited and dug out the stones from this mine in Balakhshan said that he had seen in these mines where spinels are found a stone containing water that had still not completely matured.

Good and bad qualities

Spinel comes in three varieties: a red one, called *mu'aqrabi*[4] after the scorpion; a peridot-green one,[5] and a yellow one, the best being the red variety. None possesses the properties and benefits of corundum; their only advantage is that they look like corundum in color, clarity, and brilliance.

A gem dealer told me that the varieties of spinel are five: the *mu'aqrab,* which is the deep red variety, followed by the *'atash*[6] of lesser red, followed by the *inari,* the color of the pomegranate, which it resembles, as in Persian *inar* means pomegranate; then the *niaziki,* of lesser red shade than the *inari;* then the yellow, of a color similar to yellow sapphire.

Value and prices

In general, the value of a good spinel is half that of a good corundum.

Notes

1. Literally: similar to emerald and corundum.
2. The line inside the bracket is missing from this version. It was taken from the Cairo one.
3. "Which is a town in Turkish lands" in the Cairo version.
4. In Arabic, *'aqrab* means "scorpion."
5. Literally: a green one and a peridot-green one.
6. *Mu'tash* (?) in this version.

Chapter 6: On Garnet

WE have already mentioned that the formation of garnet and spinel and their mines [location] are the same.

As for the varieties of garnet, they are four: *mazenbi,*[1] of a light red color, clear and transparent is the most expensive kind. I asked an elderly gem dealer about the reason why this variety was called by that name; he said that this stone is very similar to good ruby, and as it is priced below the value of ruby, it seems to wonder: What did I do wrong to be valued below ruby? Then, the red variety: of a strong red color, called the *ratb;* the *bahandji*[2] variety, of a reddish-black color with a slightly reddish shade coupled with a tinge of blue; the *azbazashtt,*[3] of a light yellow color. All [varieties] are similar to spinel but are of darker shades.

Properties and benefits

Only *azbazashtt* of all the garnets has the property of stopping nose bleeding when hung externally. I do not know any properties for the other garnets.

Value and prices

The value of garnets is a quarter of that of spinels or even less, depending on their color, clarity, and kind. The best variety is the *mazenbi* priced at 2 dinars per *Mithqal;* the red one is valued at half the price of *mazenbi,* the *azbazashtti* at half that of the red variety and the *bahandji* one at half that of the *azbazashtti.*

Notes

1. Literally meaning: What is my fault, what did I do wrong?
2. Or "*bahdji*"? The word was not clear in this version, it was taken from the Cairo one.
3. Also: *azbazashtti*.

Chapter 7: On Tourmaline

CAUSE of formation of the stone in its host rock

We have mentioned this earlier under chapter 5.

On the mine [location] where it is formed.

Tourmaline is found where corundum is found, in the island behind the island of Sarandeeb in the mountain known as Djabal al Rahun described earlier. Nowadays it is also found near Bijadi, these stones being of better quality than the *hindi* [Indian] ones.

Good and bad qualities

Badjadi is a stone of wine color, without brilliance except in a few specimens, these resembling ruby. It is colder and softer [1] than ruby, and if the stone is found in its host rock, it is dark and opaque; only when cut by the specialists do its color and beauty appear, and it becomes brilliant. The best is the one of strong red color, with good brillance. It does not shine unless it is foiled, and its underside is hollowed out, this being done on most stones. Only those clear stones containing a lot of moisture can show light with a flat unhollowed underside, these stones being rarely cut.

A stone that resembles tourmaline is almandine garnet, which is a deep red stone with a tinge of black, softer than tourmaline, so dark in shade that its underside needs to be hollowed out to make it thinner in order for its color to show; otherwise its clarity does not show. The best is the one with the deepest color that does not need too much hollowing out; it is less valuable than tourmaline, and it differs from it by the higher moisture content of tourmaline and its better clarity, also [by the fact that] it does not pick up any dust particles from the floor.

Characteristics of tourmaline

One of the characteristics of tourmaline is that when it is rubbed against the hair of the head or of the beard, then is placed on the floor, it picks up straw and other objects. Another property is that if female animals–speaking or nonspeaking–see this stone, they become in heat and cannot control their senses.

Properties and benefits

If 20 grains of barley worth are worn in a ring, it prevents nightmares. Also if 4 grains of barley worth are powdered and made to drink by a person suffering from ascitis,[2] the water will be eliminated at once, and the person will be cured. Ibn Hamdan[3] said that it strengthens and safeguards eyesight; it also removes veils[4] and darkness.

Value and prices

One *Mithqal* is worth 1/2 dinar. Some gem dealers consider that the varieties of garnet are five, including tourmaline in the last, fifth place after *azbazashtti*. It is priced at half the price of *azbazashti,* which we have already mentioned.

Notes

1. Meaning "less hard than ruby."
2. Ascitis is excessive water retention in the stomach.
3. Unidentified person.
4. Albuginitis or the beginning of cataract?

Chapter 8: On Diamond

CAUSE of formation of the stone in its host rock

Pliny said that diamond is a gold base stone and that it resembles more the molten substances as it cannot be crushed by any stone like other stones which can be crushed by each other; for this reason I [Pliny] compared it to the [above-mentioned] substance. Nothing can spoil[1] it except lead; this is why I said this is a gold base stone, and I say that diamond started in its host rock to become gold, as water was present in its initial constitution in its host mine. As it was heated, the diamond part in it became strongly heated and it solidified into a stone. And when the heat increased further, it expanded and thickened and became viscous, resembling mercury very closely. Then the moisture in the mineral and its dryness became balanced as a result of its formation under moderate conditions,[2] creating salt, which was then dried by water and air, thus solidifying. Then the heat increased, and the salt was able to get rid of the heat and the dryness, its hardness reaching the top of the viscous matter resembling mercury. Thus the stone was born as its hardness increased. It started initially to be gold, but drought and the presence of salt[3] prevented this, and had it formed under moderate conditions without excessive drought and with the presence of sweet rather than salty substances, it would have become gold. But as it formed under prevalent very dry and salty conditions, it did not turn into gold, but became a hard stone able to eat [scratch] all stones due to its salty and dry nature. It could break due to the presence of salt, as the salt and dryness remained in its constitution. The only substance that could harm it is lead, as it is a gold base stone, and lead can spoil gold and destroy it. Lead can destroy diamond due to its contents of sulfur, as a result of the combination of sulfur in the lead with the salt in the diamond; as when the salt in the diamond detects the smell of sulfur, it breaks into pieces and is destroyed.

Diamond became white as it solidified under moist conditions, which repelled the heat of the fire [or sun], and made the stone white. This is why diamond is a white stone.

On the mine [location] where it is formed

It is found like ruby, as pebbles in the ruby mines, found together with it, transported by rain and wind from their host rock, according to what we have described on many earlier occasions.

Yuhanna ibn Masaweih had mentioned that diamonds are found in a valley in India so deep that people could not reach its bottom; diamonds being scattered in its bed, ranging in size from the mustard seed to the barley grain. Therefore, tender meat pieces are thrown in the valley, where eagles see them and try to take them, rubbing them in the earth as they go to devour them, thus making the diamonds adhere to the meat. Then many other eagles come and fight over the meat, then carry it away, making the diamonds fall off to be picked up by these people. These eagles are well known to do that and they are constantly watched [by the locals].

Good and bad qualities

Diamond comes in two varieties: the *zayti* one and the *billawri* one, the *zayti* being the most valuable. The *billawri* variety is of a strong white color, similar to rock crystal; the *zayti* is a white color mixed with a tinge of yellow, similar to the color of oil and resembling pharaonic glass.

Some Persian gem dealers who often visit India and China looking for valuable gemstones informed me that there is a variety of diamond with such magnificent brilliance that they can project this brilliance on a nearby wall, piece of cloth, or face of a person or other object, by producing light of various colored rays, similar to a rainbow.[4] This variety of diamond is worn by Indian dignitaries for adornment purposes; they keep it jealously for themselves. As for the rest of the diamonds that do not display this dispersion effect, they are used for cutting corundum and are given to the merchants.

Characteristics of diamond

A characteristic of diamond is that all of its varieties have fixed angles; six angles, eight angles, or more or less; its angles are surrounded by surfaces of triangular shape.[5] If it breaks, it breaks only in triangular shapes even if broken into the smallest of parts.[6] Another characteristic is that it can cut every stone while it is itself extremely difficult to break; even if it is placed on an iron anvil and hit by the heaviest hammer, it does not break, but it goes into the anvil and into the hammer and breaks them both.[7] The way to break it[8] is to put it inside a piece of resin, then place it inside a wicker reed. Then it is lightly and carefully tapped with a hammer made of lead, making sure that the iron does not touch the diamond; or it is placed in a tube of lead, and is hit carefully by some hard instrument, then it can be broken.

It [diamond] is a hard stone that can eat [scratch] all other stones, as we already mentioned, and any stone that comes in contact with it will be deeply scratched. It also removes all sheen from stones when rubbed against them and can erase all engravings; while, despite what we have mentioned, it cannot be broken [or scratched] by any stone as it softens and weakens them all, scratching them deeply. The stone that can be scratched by all others while it cannot do so itself [and the only stone that affects diamond][9] is lead; thus the sages have concluded that diamond is a gold base stone, as lead can spoil diamond, as already mentioned, and not any other stone, as it can spoil gold. Another characteristic [of diamond] is that it attracts flies, and if the piece is small enough, the flies swallow it and carry it away. Ahmad said: It is said that one characteristic [of diamond] is that if a person swallows a piece of it, even the tiniest piece, it tears his intestines and kills him on the spot. A trustworthy jeweler in the *Maghreb* told me so, saying he had seen this himself. Then a group of jewelers from both the *Maghreb* and the *Mashreq* told me that many times, they placed a small piece of diamond in their mouth and swallowed it without being harmed. I searched for the truth in this matter and ascertained that both versions were true, as diamonds are found in a valley full of snakes, with some of their saliva dribbling on the stones. Whoever swallows a diamond covered with this hard saliva is killed on the spot, and if the stone is placed in the mouth, without being swallowed, it causes irritation and itching of the gums. Stones that are free of snakes' saliva either because of their initial location or for some other reason do not cause any of the above-mentioned harm. For this reason, it is thought that [the above] is an inherent characteristic of diamond, while the truth is what we mentioned. Another characteristic is what Aristo mentioned, that there is such a great affinity between it [diamond] and gold that it sticks to gold wherever it is, and a piece can sometimes be embedded in it.[10] This is known to gold craftsmen, as sometimes a small piece [of diamond] falls under the files while they are working the gold, spoiling them or eating them away. Another characteristic is that it can drill holes into pearls, corundum, emerald, and other stones, as well as any stone that cannot be worked with iron, as easily as iron can drill wood; this [is done] by placing a piece of iron on one end of a piece of wood, on top of which a stone of the size wanted to make the appropriate hole is placed; then it is used for making this hole, which it does in no time. Diamond is cold and hard in the fourth degree.

Properties and benefits of diamond

As mentioned by Aristo and tested and found to be true, if a person suffers from a stone in the bladder and the urinary tract, and a piece of diamond is stuck firmly with mastic[11] on a probe made of copper or silver, then this catheter is introduced until it reaches it, the diamond will shatter this stone to pieces. Ahmad ibn Abi Khalid, known as Ibn al Djazzar,[12] said in his *Book on Stones* that he had treated the servant Wassif, in charge of transmitting people's complaints to the judge, of a big stone that he had refused to remove by iron [surgery] with this method. When we did the above procedure, the stone was shattered to pieces, and what was left from it came out easily through the urine.

Another property mentioned by Aristo in his *Book on Stones* is that diamond is useful against strong stomach pains and dysfunctions if hung externally over the abdomen. It is also mentioned in Ibn Saharbakht's[13] book that it is a very hard, cold, and dry stone in the fourth degree; it is also said that it is hot and dry; for this reason it can polish teeth, whiten, and clean them. If it is worn by a newborn at birth, it protects him from fear and epilepsy; and if it is used in the mouth,[14] it fixes teeth [firmly in the gum] but could kill if it retained on it some of the snake's venom, as it is found in the valley of the snakes.

Value and prices

The common average price of diamond is 2 dinars per carat. Ya'qub ibn Ishaq al Kindi had mentioned in his *Book on Stones* that he had seen sizes varying from the mustard seed to the walnut and that he had not seen any larger than that. He also mentioned that the most expensive stones he had seen in Baghdad are sold at 80 dinars the *Mithqal.* He said that if a rare large stone of 1/2 *Mithqal* suitable for a finger-ring is found, its value is double that of the mustard seed size, three times the *filfila* [seed] and four and five times more. And he had said that the cheapest he had seen in Baghdad sold at 15 dinars the *Mithqal.*

Notes

1. Literally: destroy.
2. Literallly: during gentle cooking.
3. Literally: water, but salt is mentioned in the Cairo version and this is the correct meaning.
4. This is a reference to fire in diamond.

5. Meaning the triangular faces of the octahedron.
6. This is a reference to cleavage in diamond.
7. This is indeed how diamond was tested to ascertain its genuineness until at least the fifteenth century.
8. Meaning the way to cleave it.
9. This sentence was taken from the Cairo version to complete the meaning.
10. Any piece of gold can be an inclusion in diamond.
11. A resin from the *Pistachia Lentiscus*.
12. Ibn al Djazzar, physician born in Qairuwan, Tunis, around A.H. 285.
13. Ibn Saharbakht, circa tenth-century physician.
14. Probably meaning by massaging the gums.

Chapter 9: On Moonstone Cat's Eye

CAUSE of formation of the stone in its host rock

This stone started in its host rock to be a corundum but external factors prevented it from becoming a corundum, as they prevented diamond and other stones that started to be corundum, according to increase or decrease in the four elements. But the degree of moisture and clarity is stronger in diamond and the other aforementioned stones than in the ones that started out to be corundum.

On the mine [location] where it is formed

This stone is found in the same location as corundum and diamond, as was mentioned earlier.

Good and bad qualities

This stone is a strange one, as the overwhelming color is white with an astonishing radiance [coupled] with a delicate transparent clarity, but a bluish spot can be seen inside it, similar to a cat's eye catching the light in its moving pupil; this spot moving constantly. If the stone is moved, it moves in the opposite direction, in such a way that if it is moved to the right, the spot moves to the left and vice versa, likewise upwards and downwards; it is exactly like the eye of a cat, and for this reason it was called cat's eye. If the stone is broken or cut into tiny pieces, this spot will appear in every single one of these pieces. The best quality is the strong, white, transparent one with a clear spot showing little movement but great brightness. If the stone is held still so that its brightness appears, water similar to moving waves is seen inside it, so clear that it illuminates its surroundings. If it is of top quality, the movement of its waves increases, giving the impression that it contains water, but there is no water in it; rather it is a transparent gemstone, showing this [characteristic].

The best stone is the transparent one with beautiful appearance and large size, which increase its value and the demand for it, as is the case for other stones.

Properties and benefits

I did not find any mention of this stone in any book on stones, whether old or new. I think it has come recently in people's hands; its well-known property amongst gem experts is that it protects its wearer from evil eyes and evil spirits, whether known or hidden, human or occult.

I also report the following based on what trustworthy jewelers who visited India and roamed there looking for unusual properties of stones and studying their curious nature and their secrets told me, namely, that this stone possesses all the properties of the *bahraman* ruby, in addition to two great benefits: The first is that it safeguards the money of its wearer and protects him from misfortunes and catastrophes, whatever they may be; the second is that if it was worn by a man and he participated in a war where his side was defeated and there was no way out to save himself, if this man lay amongst the killed, all his enemies who come near him see him as dead, covered with blood, thus are repelled, and no person approaches him.

A jeweler who visited India also told me that he had seen this stone in Ma'abar[1] and that it was worshippped there as idols were worshipped. He said it was more valuable there than in Arab lands, as they appreciated it and esteemed it more. He also said that he saw such a stone being sold in Ma'abar for 150 dinars, which would hardly be worth 1/10th of this price elsewhere, as they knew of its secret properties that are unknown to other people and had tested same.

Value and prices

The value of this stone depends on demand for it and on people's knowledge of its properties. But in Arab lands, the *Mithqal* is worth only 5 dinars, increasing according to demand and knowledge of its properties, as already mentioned. In India and the neighboring foreign lands, it is many times more expensive than in the Arab lands, contrary to most other stones, the value of which decreases with the closeness of its mine and increases with its far distance. I was informed by a man from Ghaznah that he saw a stone sold in his country for 700 dinars.

A man from 'Aden whom I trust entirely told me that a man called Ismail al Salameh died there in the year 641 [A.H.] and he left his heirs a bag containing a moonstone cat's eye weighing approximately 1 *Mithqal,* together with a paper showing that he had bought this stone in Ma'abar for 700 *Fili,* the *Fili* being like the gold *Ya'qubi* dinar there, worth 5 *Maliki* dinars, which is equivalent to 1 1/2 Egyptian *Dinars.* He said: Ibn Rasul,[2] the ruler of Yemen, bought this stone from the heirs of the above-mentioned deceased person for the price of 7,000 *Maliki* dinars, and it was thought that this price was unfair to them. Apparently the stone showed such a clarity that it used to emit rays and shimmer, as described earlier.

Notes

1. A province in India.
2. The *Banu* (sons of) Rasul dynasty ruled Yemen from A.D. 1229 to 1454.

Chapter 10: On Bezoar

CAUSE of formation of the stone in its host rock

Ahmad said: The bezoar stones available nowadays are of two kinds: one animal, the other mineral. As for the mineral variety, I have seen its mine myself, located on the borders between Djazirat Ibn 'Issa[1] and Mosul,[2] where it is found in large quantites. It is also found in large stones that are made into knife hilts and other objects; one piece could reach 2 ounces or more in weight. This variety is white with spots of yellow and other colors, and it is of no benefit whatsoever against poisons. This stone is soft and can easily be filed, its streak powder being white; it is easily scratched and does not possess any property or benefit except that if it is dissolved in water and spread over a swollen ecchymosis of blue or black color, it cures it; the swelling and the pain are removed, and the color returns to normal in no time, not exceeding a day and a night. I have tried this myself more than once and found it true as I told you. The mineralogist I mentioned under the emerald chapter told me that big bezoar stones of whitish color or more than that are found in the emerald mines in Egypt in very large quantities. What is available now from the mineral bezoars is a variety brought from Soghd,[3] small yellow stones of uniform color or showing specks of different colors, the powder of which is of little benefit against scorpion bites only. As for the animal bezoar, which is the purpose of our discussion in this chapter, it is a light brittle stone of yellow or dusty color showing small spots. It is constituted of thin layers, one covering the other, and can only be this way. It can be scratched quickly, and its streak powder is whitish. The biggest stones weigh from 1 to 7 *Mithqals;* they are brought from Persia at the border with China. The animal where it is found is the local deer. The animal where bezoars are found likes to eat deadly venomous snakes, especially their young, this food constituting most of his meals; he searches for them, brings them out wherever they are, and eats them.

People have different opinions as to in which place this animal bezoar is formed: The first opinion is that it is formed in his eyes as a disease, as when the deer has eaten too many young snakes, his body starts itching all over due to their poison. So he goes to a pond and goes into the water, keeping his head out until he is completely immersed in the water except for his pupils. At that time, moist vapors arise from his body and reach his eyes, then go out through his left and right eyelids near his nose and condense into water. When the air hits it, it solidifies and becomes stone, remaining stuck on the hairs near his nose. Then the same thing is repeated, with more vapor rising and turning into water, then falling through that place in its eyes and solidifying over the previously-formed stone as the latter had solidified, and so on until either the stone becomes heavy and falls off by itself, or when he feels its heaviness, the animal rubs it against a stone or the trunk of a tree, where it falls, is found and taken.

The great learned Emir Sunkur ibn Qilidj[4]–may God raise his standing and protect his glory–told me that he had seen with his own eyes during some of his hunting trips in the borderlands between Aleppo[5] and the Rum[6] lands in a place called Mar'ash[7] and its surroundings, deer that immerse themselves in ponds until only their eyelids show. When he enquired about the reason for this, he was told what we have explained here above about eating snakes and the bezoar stones coming out of their eyes. For this reason, he said, I strived to hunt a deer and found in his eyelids humid compact substances, some the size of a chick pea; others the size of a fava bean. I took out some and gathered them. At that stage, he took them out and showed me these stones. I saw that they had a blackish tint and were still not completely solidified, humid like wax. I asked him, did you test it against any poison? He said no, but I ascertained that deer eat baby snakes. Then the emir bared his arm, where a silk rag was tied holding a small round purse, the size of a large hazelnut or slightly smaller, containing a stone. He said that king Ashraf[8]–God bless his soul–had given it to him, and he said that he knew of only two other similar stones, one remaining in the possession of king Ashraf, the other one with the Caliph–God bless his soul–in Baghdad. I asked him about the property of this stone, and he said that if a person bitten by some poisonous animal puts the stone on the place of the bite, the stone sticks to the place and extracts the poison by suction, the stone remaining stuck until all the poison is removed, and the person is recovered; then only does it fall off. He mentioned that he had tested this property with a person who was made to drink poison by placing it in his mouth. The stone attracted the poison out through vomiting. He did not know its name or whether it was mineral or animal.

The second opinion is that this stone is formed in the heart of this animal. The animal is hunted for this purpose and is slaughtered, and the stone is extracted from his heart. This is only the opinion of some late physicians of Egypt, and it is wrong.

The third opinion is that the stone is formed in the gallbladder of this animal, like many stones that are formed in the gallbladder of many animals. The deer is hunted, slaughtered, and the bezoar taken out of its gallbladder. In my opinion, this is the truth of the matter, and this is similar to the bezoar brought from China.

A scientifically minded trustworthy elderly gem expert known for his honesty told me that if one licks a genuine one of these recently discovered stones, he finds it to have a strong bitter taste. This person told me that he had broken a bezoar stone, finding inside it an herb around which the stone started forming in its host rock.

On the mine [location] where it is formed

As for the animal variety, you have gathered from what we have already mentioned in which part of the animal the stone is formed and in which countries this animal is found. Likewise for the mineral variety; we have already mentioned where it is formed, and we have described it.

Good and bad qualities

The pure, genuine, useful bezoar found nowadays is the animal variety and not the mineral one, which is yellow, brittle, and light in weight, of dusty color and showing small, light spots of white, with a bitter taste. The best mineral variety is the Chinese one, which are small lemon-yellow stones with green stripes. Aristo had mentioned the mineral variety, saying that there were many other [varieties], including yellow, dusty, one with a reddish tint, and one with a whitish tint. I was told that the best is the yellow one. Animal bezoars are often faked and cheated, thus small stones are manufactured from layers of different masses, made to resemble animal bezoar. It is often cheated, and the test that differentiates it from [genuine] animal bezoar is that the manufactured one is dusty, dark, and plain in color, while the genuine one is yellow or dusty with a yellowish tint, spotted with small specks like freckles, with much thinner layers than those of the fake manufactured one. It is also lighter or similar in weight and more brittle than the manufactured one, of pure milky white streak powder. The correct experiment is to test it with poison.

A dignitary from Morocco who attended the *Madjlis*[9] of its great king Ya'qub ibn Yusuf ibn 'Abdul Mu'min[10] told me that one day, a parcel full of stones was brought, and the king called the master physicians and jewelers' supervisors and told them that these were bezoar stones that had been gathered in our land. He requested that these stones be tested and, if found genuine, be distributed to market supervisors and clan leaders all over the city; this was to be well advertised. This was done for [religious] credit purposes, [as these stones were] to be used by all those who needed them as necessary, for curing bites by poisonous animals. He said: The stones were counted; they were 280. The doctors came; snakes were caught and made to sting chickens. The stones were filed and their powder fed to the chickens. The stones that were given to chickens who survived proved to be genuine, while those given to chickens who died were not. Only sixty stones were found to be genuine, while the rest were fake and cheated. He [the king] kept some of these genuine stones and ordered the rest distributed, as mentioned.

Characteristics of animal bezoar

If it is placed near a scorpion, it renders its sting harmless, and even if it did sting, its poison will not be effective. By nature, the mouth and throat of snakes are larger than their bellies. If two grains of barley worth [of this stone] are powdered and dissolved in water, then are poured in the mouths of snakes and vipers, it chokes them and kills them. This, together with what was said previously, is how genuine bezoar is distinguished from the fake one. Another property is that if it is placed with rough substances that come in direct contact with its body, scratching it, its appearance will be altered, and it will be so scratched that it will become rough, lose its freshness, and change so much that it will almost be unrecognizable.

I owned a genuine animal bezoar stone, which I placed in a bag containing gold dinars. Then I left on a faraway trip, and when I reached my destination, I opened the bag and took out the bezoar stone, which I hardly recognized. I even thought that it had been substituted as all its characteristics had changed. Then I weighed it and found it to be lighter than it had been. My suspicions increased, but as I had no one to suspect, I was astonished and wondered about the matter. Then I placed it in a small bag after wrapping it with silk and left it there for some time. When I took it out again, I found the stone I had known previously, after the ugly appearance that it had acquired from contact with the roughness of the gold was gone, and it had recovered its original characteristics. But its weight had decreased from contact with the gold in the bag. Later on, as I was talking about bezoar with a shrewd jeweler, he informed me of the stone's characteristic of being altered upon contact with rough substances, which I confirmed by relating my own experience.

Properties and benefits of animal bezoar

Bezoar is a foreign Persian name; the origin of the word in Persian is *Bak Zahr*, *bak* in Arabic meaning cleanliness and *zahr* meaning poison. Thus, it means cleanser of poison from the body. When it was arabicized, the "k"[*kaf*] letter was dropped and the word became *Bazahr*. Its specific characteristic is that it is beneficial against poison, any kind of poison, deadly or not, whether animal, plant, or mineral, also from hot and cold poisons and bites by poisonous animals. If 4 to 12 grains of barley worth of this stone are powdered or scraped with a file or filed on the whetstone with olive[11] oil or water and are drunk, they can extract the poison from the body of the victim through perspiration; thus he escapes from death. It does that on account of its constitution and inherent properties; it is a noble, valuable stone, and no other stone whatsoever can repel poisons like it. Another property is that if it is powdered and sprinkled over stings or bites from poisonous animals, it extracts the poison and neutralizes it. Another property is mentioned by Ibn Djumai'[12] in his book called *Al Irshad li Masalih al Anfas wal Adjsad* [Instructions in the Interest of Souls and Bodies], who said that animal bezoars that are found in the hearts of deer are the best of all bezoar stones in all aspects.

If it is ground on the whetstone with water and half a *daneq*[13] is given daily to healthy persons as a preventive measure, the person will [be able to] resist deadly poisons and will be immunized against their harmful effect, not fearing ill effects nor adverse reactions in the body fluids [four unclear words][14]. It does no harm to people with fever or those that are ill, as its action comes from within its constitution. He said: This stone contains little heat and is beneficial against all kinds of animal and plant poisons, hot or cold; and if a stone weighing 1 grain of barley is set in a ring and placed on bites by scorpions, poisonous insects, winged or not, wasps or blister beetles, it is of great benefit.

Another property is that if it is powdered and sprinkled immediately on bites by earthy[15] animals, it attracts the poison by suction and neutralizes it, and if the wound occurred before it is treated with the medicine then is sprinkled with the powder of this stone, [the person] will be cured.

Another property mentioned by the ancient sages is that if bezoar is set in a ring of gold, and the picture of a scorpion is engraved on it while the moon is in Scorpio, and Scorpio is one of the poles of the ascendant; then stamps of frankincense made into pills are printed with [the engraving on] this stone, and a person bitten by a scorpion swallows one of these pills stamped with the bezoar stone, he will not be affected by the sting, and he will recover. Ahmad ibn Yusuf,[16] ibn Tulun's[17] secretary and commentator[18] of Ptolemy's book *Al Thamara* [The Fruit] said that he had tried the above, and it proved to be true. He had also used the stamp on substances other than frankincense, and the result was the same as when frankincense was used.

Another property is that if a person possesses a finger-ring [of bezoar], and either he or someone else was bitten or drank poison, if this ring is placed in the mouth of the poisoned person to suck at, it will be beneficial to him, even if the stone does not lose some of its size.

The author of this book said: One of the wonders of God Almighty's Creation with regards to animals is that antipoisonous stones are extracted from deer that are found in the above-mentioned locations in China, as we have already mentioned; and a deadly poison can be extracted from deer of both the *Mashreq* and the *Maghreb*, as the knot at the end of the deer's tail of the *Mashreq* contains a deadly poison. And I think that this is also found in bezoar-producing deer; this poison kills by drugging and freezing the blood, and I have seen many kings both in the *Maghreb* and the *Mashreq* order hunters to bring it to them to be kept in the royal treasury.

One of the most astonishing wonders of God's Creation in His secret Wisdom is embodied in this animal, who feeds on the deadly poison of each and every animal, this poison running in his body without harming him, and combining in it the poison itself at the end of his tail and its antidote in another part of his body. Despite the above, its meat is considered to be a healthy food of no ill effect whatsoever, and if any place is fumigated with his horn, no poisonous animal will approach it. This has been tested. God be blessed, He Who knows all, controls everything and can do anything.

A hunter who used to hunt deer in the *Maghreb* and remove the knots off their tails to give to the king told me a strange story about deer. He said he came upon a big fat deer while [hunting] in Africa,[19] near the town of Qafsa,[20] accompanied by his friend. They removed its knot, put it aside, and proceeded to dry the meat of the deer in order to carry it dry to town, as was customary. He said: We cooked a big part of this delicious fatty meat, which was covered with a lot of fat, and we had ourselves a banquet eating it. Then we took a nap under the shade of the tree where we hung the meat to dry, and when we awoke, we were hungry, and we discovered that the meat had dried out in a way that we had never known it to dry during the time we slept. We carried it to the city. The day we slept was Tuesday morning, but we found the people coming out of the Friday prayers. As the place where we had slept was only a one day's walking distance away from that town, we realized that we had slept approximately three whole days and nights in one go. This same hunter told me that the above-mentioned knot kills by drugging, in a sleeplike manner, whereby the person who eats it sleeps without feeling any pain at all, a continuous sleep from which he does not awake, this poison killing by freezing the blood of the heart. Thus the person dies totally painlessly. The author of this book said: I have often eaten this meat, which the above hunter who told me this story and other hunters gave me. It used to be given to me already dried, and I would eat from it. It is a good tasty meat.

Many stones found in the bellies of animals have strange properties, for example, Dioscouridis mentioned that if the chick of the swallow is taken during the full moon[21] provided it is the first of his chicks, and its craw is split open, two stones are found, one of a uniform color and the other of different colors. If these stones are placed in a leather bag made of veal or deer skin before they are soiled with sand [or dust?], and then are placed on the upper arm or the neck of an epileptic, it will do him good. He said he had done this many times and cured epileptics completely. Alexander[22] had also mentioned this stone, saying that two stones are found in the craws of baby swallows and in their nests; either both white or one white and the other red. If the red one is hung on an epileptic person, he will be cured; and if the white one is placed on a person in an epileptic fit, he will come out of it; and if it is hung[23] on him, he will not suffer from any epileptic fits. Al Tabari[24] said that the jaundice bead was well known, of yellow color; and if it is hung on a person suffering from jaundice, it is greatly beneficial to him. This stone is also found in the nests of baby swallows. He and other scientists knowledgeable of the properties (of stones) said that baby swallows often have jaundice, and if they do, adult birds fetch this bead and place it in their midst, and the disease is repelled. If one looks for this bead in their nests without success, the following trick is done to obtain it: While the adult swallows are away, the faces of the baby birds are coated with turmeric dissolved in water. When the swallows return and see this, they think that their chicks have jaundice, so they fetch the bead and place it in their nests, from where it can then be taken.

Aristo had mentioned in his *Book on Stones* that stones are found in the craws of roosters, some whitish, others of strong white color. If you can obtain one of those stones and hang it on an insane person, he will return to sanity; and if a young man wears it on his thigh, his prestige will be enhanced, and his copulatory powers will increase. It also repels from him the devil and all evil, and is beneficial for young boys who are afraid in their sleep and those who grind their teeth. The gallbladder of cows is also of benefit as it contains something resembling a stone. This is taken and given to sniff by persons suffering from weak vision or from potential cataract in the eye, and it is also given to sniff to persons who suffer strong epileptic fits. An amount equal to a lentil in size diluted in water where chard stems have been squeezed [is given], and it will be of great benefit to them.

Ahmad ibn Abi Khalid, known as Ibn al Djazzar, said in his *Book on Stones*: If stones formed in the kidneys and bladders of people are powdered and mixed with collyriums, they are of substantial benefit to whites of the eyes. The stones used for bringing rain are a well-known true fact in the *Mashreq.* A person whose honesty I do not doubt told me a well-known anecdote in this respect; namely, that he had seen in Bilad al Fors [Persia], between Bokhara and Samarqand, in the military camp of the late king, sultan of the foreigners, 'Ala al Din[25] Mohammad Khwarizm-shah,[26] a Turk who works[27] with the stones that we are describing, and he could bring about pouring rain whenever the sultan so wished.

He said: In China and Persia, there is a kind of bird called *surhaf* which means in their language "of beautiful water" as *surh* means most beautiful and *af* water. He went on: This bird looks like a large red-feathered goose. He said: This is the bird called in Egypt *bashmur,* which is abundant there, its feathers being used to decorate boats. He said: In China and the surrounding lands, as well as in many Persian lands, this bird builds its nest in marshes during the torrential rainy season, and if the water dries up, he leaves his nest and digs about two cubits underneath, where a single stone of dusty color with white and red spots, the size of a large egg, is found. This stone is a soft one,[28] and the softer it is, the better it is. As many stones as possible are gathered and taken to the king's coffer, where they are placed in a chest under lock and key, guarded by a reliable person. If the king goes on some trip during the summer, and he is bothered by the heat and the dust of the road during a season of drought, or for any other reason that calls for abundant water and cool weather, he orders the person in charge of the stones to use them [for this purpose].

A person from Ghaznah who had witnessed how these stones are used in the camp of the above-mentioned sultan said: I saw a Turkish sheikh who came, and a tent was erected to shield him from the people while the man in charge of the stones and myself were present. He said: The top of the tent was left open to the sky, without any cover. A large vessel containing water was placed in his hands, then he took three thick sticks, placed one on the right side of the vessel, the other on the left side, and placed the third one in width over the vessel, with its two ends between the two raised sticks. Then he took out a thin snake of the same color as the stones, that is, dusty with reddish and whitish spots. He tied a thread on its tail and hung it upside down on the middle stick; its head about 2 cubits from the water, then he took two of the above-mentioned stones from the man in charge of them, placed them in the water and took them out, rubbed them slightly one against the other, then threw them in the water, then [took them out] and rubbed them again, then threw them. He did this seven times, then he took a little of this water and sprinkled it on the floor. He said: The person who was doing this act was bareheaded, as his condition to perform this deed was to be bareheaded; his hair was loose, and his face was distorted as if angry. All during this time, he was pointing with his head towards the sky and saying words invoking the rain. He did this for two hours, and soon the sky darkened, and heavy pouring rain came down. The storyteller said he witnessed this act to see the strange wonders and secrets of God's Creations, arriving when the sky was clear and departing only under pouring rain and in the mud and this during the arid season. He said: The Turkish sheikh who did this act told me that everytime he did it, a misfortune happened to his family: Death of a son, loss of money, or some other mishap. But the king would always compensate his loss. Only people specialized in this carried out this ceremony.

This storyteller as well as other trustworthy scientists, merchants, and dignitaries from Persia told me the following, which is well known in their country: The above-mentioned foreign sultan Khwarizmshah wanted to visit the Turkish lands lying on the borders with China. When he approached them, he was faced with such heavy cold, rains, and snows that they almost destroyed his army. As it was not the season for such heavy rain and cold, having checked the time of the arid season due to the extreme cold of their country, he realized that this was being done with the above-mentioned stones, so he sent some of his soldiers to roam and search the mountains. They brought him two men they found working with these stones, whom he wrapped in two black fleece, and buried alive. The rain, snow, and extreme cold stopped that same day. He said: He [the king]ordered that the same should be done to anybody found doing what we mentioned hereabove, otherwise the effect of his act would last a long period of time. If this act was done in two places, then the cold and snow became so hard that they were unbearable. God be praised, He who placed the Secrets of His Wisdom in His Creations, that cannot be understood except by those of great minds. He said: And how many miracles in heavens and earth are witnessed by them, but still they refute them.

A merchant from Persia who often visited Turkish lands told me that he had gone to one of the Turkish towns called Tahan. He said he found that snakes were their most numerous inhabitants. He said: Huge snakes fall from the sky there, and he bought the skull of a snake for 55 dinars, out of which he made three knife hilts, worth 100 dinars each, which he sold to the king of Persia for 500 dinars. He said: Its characteristic is that if it is placed near some poisoned food or drink, it sweats profusely. He said: The effect of its powder on these bones is obvious. He said: And if some bone comes out of it, and a dog stricken with rabies or a donkey sees it, he is mesmerized on the spot and falls unconscious on the ground, [remaining so] two to three days. He said: I tried this myself.

Value of bezoar and prices

Mineral bezoar found today in the hands of people is of no value worth mentioning as it lacks the properties and benefits found in animal bezoars. As for animal bezoars, the manufactured and fake ones are more abundant than the genuine ones in peoples' hands.

I was visiting the shop of a jeweler who was [also] a gem expert from *al Andalus* in Alexandria when a foreign merchant came and took out eighteen bezoar stones supposed to be animal bezoars. He gave them to the market broker, who showed them to the market supervisor, who did not single out any stone, confirming that they were all animal bezoars. When the stones came into our hands and the jeweler in whose shop I was saw them, he picked two stones that he showed me, telling me that out of the whole lot, only these were genuine bezoars, the others being manufactured or falsified. He proved the truth of his words by pointing out characteristic features of the genuine and the manufactured stones, which are apparent to the intelligent, observant, cunning person, as manufactured stones cannot deceive the bright intelligent person. We asked the foreign jeweler to sell us only these two stones, which he refused, saying that he would sell only the whole lot as he had bought it; and when we confronted him, and he still would not budge, we took him aside and told him that the stones were all manufactured and fake except for two of them, which we proved to him. He broke down then and admitted that he had bought them as such, and told us to take one stone if we so wished. I bought one of the two [genuine] stones for the value of 1 dinar per *Mithqal,* and he sold the rest at this [same] price. I also saw in the market in Cairo–may God protect her–many fake manufactured stones being sold as animal bezoars for the value of 1 dinar per *Mithqal,* and if these stones were tested on the throats of snakes, they would not kill them; or on the scorpion's sting or any other aforementioned test, they would not be effective. A genuine stone could fall between the lot, which is sold at the same price as the fake ones for lack of experience and lack of testing; thus this price became the standard price for the reason we mentioned above.

Notes

1. Djazirat ibn 'Omar in the Cairo version.
2. Mosul, Iraq.
3. Soghdiana.
4. Seif al Din Qilidj in the Cairo version, one of the emirs of South Lebanon.
5. Aleppo, Syria, in this version.
6. Literally: *Rum* means "Romans."
7. Mar'ash, Syria.
8. Ashraf Musa ibn al Zahir, grandson of Salah al Din.
9. Council.
10. A.D. 1184-1191, Muwahhid ruler of Morocco.
11. Word not clear in this version. "Olive" was taken from the Cairo one.
12. Egyptian Jewish physician of Salah al Din.
13. *Daneq* is a unit of weight, some say it is equivalent to 1/6 of a *Dirhem,* others to 1/8th.
14. Rest of sentence illegible.
15. Meaning serpents, vipers, and all animals that crawl.
16. Ahmad ibn Yusuf (died c. A.D. 912). Mathematician and secretay of the Tulunids who ruled Egypt from A.D. 868 to 905.
17. Ibn Tulun, founder of the Tulunid dynasty in Egypt and Syria, which lasted from A.D. 868 until 905.
18. Literally: the person who explained.
19. Meaning Tunisia here.
20. Qafsa, Tunisia, at that time.
21. Written "full sea," obviously this should be "full moon."
22. Alexander the Great of Macedonia.
23. Meaning as an amulet or a talisman.
24. Al Tabari, tenth-century physician.

25. Word unclear in this version. "Al Din" is taken from the Cairo one.
27. Literally: who could do some magic, weird, deed.
26. The Khwarizmshah are a Turkish dynasty who ruled for over one hundred years (eleventh-twelfth century A.D.)
28. Literally: can easily be filed.
29. Spain.

Chapter 11: On Turquoise

CAUSE of formation of the stone in its host rock

Turquoise is a coppery stone, formed by copper vapors rising from their host rock according to what we shall mention later on the formation of other coppery stones.

On the mine [location] where it is formed

Turquoise is brought from a mine in a mountain of the mountains of Nisabur,[1] from which it is taken to many other countries. There is a variety found in Nishawar [or Peshawar?], but the Nisabur variety is better.

Good and bad qualities

There are two varieties of turquoise: the *bushaqi* and the *qadjandji.* The pure genuine one is the *bushaqi,* the best kind being the pure blue one of radiant clear color, high luster and even distribution of color. It is found mostly as small stones. Al Kindi had mentioned that he had seen a turquoise stone weighing 1 1/2 ounces.

Characteristics of turquoise

One characteristic is that it is a stone the color of which clears as the sky clears and becomes cloudy with the sky's cloudiness. Aristo had mentioned that every stone that changes color is bad for its wearer. Another characteristic is that if grease comes into contact with it, it ruins it and changes its color. Likewise, perspiration ruins it and dulls its color completely, which I have tested myself. Also, if musk comes into contact with it, it ruins the stone, removes its color, and takes away its beauty.

Benefits of the stone

It clears the vision if one looks at it. It is also beneficial for the eyes if powdered and mixed with collyriums. Turquoise is also drunk for stomach pains. It strengthens drooping eyelids and lessens protrusion of the eyes and skin and is good for veils obscuring the vision. If powdered and drunk, it is beneficial against scorpions' bites. It is a cold and hard stone.

Another property is taken from what Aristo wrote to Alexander on managing the kingdom [?], the last words in the above-mentioned epistle being, word for word: Turquoise is a stone that foreign kings still use for adornment purposes and enjoy collecting. Its great property is that it repels murder from its holder, and it was never seen in the hands of a murdered man. If it is powdered and drunk, it is useful against bites from scorpions and harmful poisonous creatures.

Value and prices

Turquoise is mostly found as small stones, as we have mentioned earlier. Stones differ a lot in quality. Stones could be worth 1 dinar or 1 *Dirhem* even it they have the same or almost the same weight. Its constitution is as we have mentioned, when we described its good and bad qualities. The Berber emirs of Morocco and their leaders like to acquire this stone and pay high prices for it, as much as 10 Moroccan dinars per stone. They decorate their swords with it and wear it a lot in finger-rings. People pretend it is used in alchemy, as al Djazzar pretended and stated in his *Book on Stones*. But this is not true; they pay high prices for it due to what is said about its property of repelling murder. God knows best.

Notes

1. Nishapur, Iran.

Chapter 12: On Cornelian

CAUSE of formation of the stone in its host rock

We have already mentioned under chapter 5 the cause of formation of cornelian and what the sages said about it. Thus, it is not necessary to repeat it here.

On the mine [location] where it is formed

Cornelian is brought from Yemen from mines in 'Aden and Sana'a. It is also brought from Mecca–may God honor her–and cornelian is also brought from the shores of the sea of Ruma and from Mecca it is brought to other countries.

There are five varieties of cornelian: red, *ratbi* which is red with a shade of yellow, blue, black, and white. The best is the red one, followed by the *ratbi,* then in accordance with the [above-mentioned] order until the last one.

Properties and benefits of the stone

Cornelian is hot and hard; it has three properties: The first property is that if the red variety is worn as an amulet or a talisman it stills fear during battles. The second property is that if the second variety, the color of which resembles the color of meat to which salt has been added, showing white banding, is worn in a ring, it stops hemorrhage from any part of the body, especially in women with long periods. The third property is that if any of its varieties is rubbed against the teeth, it removes their rust and whitens them; it also removes cavities and prevents bleeding of the gums. Calcined cornelian fixes loose teeth and makes them firm.

Value and prices

Cornelian is set in rings, each ring being sold at 4 *Dirhems,* or used to make knife hilts, each selling at 1 dinar or less. The price of a good engraved stone is 1 *Dirhem.*

The above prices all apply to the red one as it is the top variety, and the *ratbi* which is lower in price. As for the rest of its varieties, they are of no particular value.

Chapter 13: On Agate

CAUSE of formation of the stone in its host rock

We have described the cause of formation of agate under that of the formation of cornelian; thus it is not necessary to repeat it here.

On the mine [location] where it is formed

Agate is found in Yemen in cornelian mines; some is also brought from China.

Good and bad qualities

Agate comes in many varieties, including: *baqarawi*,[1] *ghurawi, farissi, habashi, 'assali*,[2] and *mu'arraq*. The *baqarawi* is a stone composed of three layers: a red opaque layer, followed by a white opaque later, followed by a transparent crystal layer. The best is the one of even thickness and beauty of banding, free of coarseness, ugly banding, and defects. As to the *habashi* variety, it is banded; its two upper and lower layers are black as jet, and the middle one is strong white. The best is the one of even layers, as described previously, as to the remaining varieties the best is the one with the highest luster and even banding. There is no stone harder than agate; it is very resistant to treatment.[3] It improves when cooked in oil, and if it is polished by hand with honey, it becomes bright and radiant.

Characteristics of agate

The philosophers mentioned that the name "agate" was derived from fear because it arouses fear in the heart. They also said: Whoever wears it in a necklace [or amulet or talisman] or in a ring will find his problems increased, see frightening nightmares, and will argue frequently with people. For this reason people in Yemen and their kings of Himyar[4] do not wear this stone and do not keep it in their coffers; they do not use it at all. Only ignorant people who do not know this property of agate wear or use this stone. People in China hate to work the agate mines, these stones being smuggled and sold to countries outside[5] China by gypsies. If it is hung on a child, he will salivate excessively.

Benefits of the stone

One benefit that Lianos of Antioch mentioned in his *Book on Stones* is that if agate is wrapped with hair of a woman who is having prolonged contractions that might be dangerous, and it is worn by her; she will give birth on the spot. Also, if agate is placed near a woman who has just given birth, it will protect her from postnatal jaundice and ease all her pains. Another property is that it heals festering wounds and blood discharges. Another property is that if the different varieties of corundums are polished with powdered agate, it improves their appearance and brings out their radiance, light, and brilliance, like no other stone can do. Agate is cold and hard in the second [degree].

Value and prices

A good quality bead weighing 1 *Mithqal* is worth 2 *Naqara Dirhems.*

Notes

1. From *baqara*, cow.
2. Word unclear in this version. '*Assali* (the color of honey) was taken from the Cairo one.
3. Meaning it is so tough and resistant that it cannot be worked easily.
4. The kings of Himyar ruled Yemen for several centuries before the Muslim conquest, until the beginning of the sixth century A.D.
5. Literally: "To China," but this is obviously an error; "outside China," taken from the Cairo version is the correct wording.

Chapter 14: On Magnetite (Lodestone)

CAUSE of formation of the stone in its host rock

In his *Book on Stones*, Aristo mentioned that all magnetic stones started in their mines to be iron, but under the effect of heat and drought, they became dry, hard, and compact stones, becoming so due to the strong heat rising from their mine, its little moisture and total drought. Thus, they became black stones of the same nature as iron, attracting it due to the original affinity, natural attraction, and compatibility between the two by nature. Iron became so obedient to this stone that if thin pieces, like nails, are taken and firmly planted in the ground, then this stone is placed on one of them, it draws others to it, where they get stuck to the one stuck on the magnetite stone, making people think they are in a specific order.

On the mine [location] where it is formed

This stone is found in a mountain above the coast between the sea of Hidjaz and Yemen; it is also found in Sana'a, Yemen. Aristo had mentioned that there is a mountain in the sea, and if ships come near that mountain, not a single iron piece is left in them; they fly off from the bottom of the ship like birds; and if a nail was bored firmly for some reason, it comes out and gets stuck to the magnetite stone. For this reason, the ships navigating in this sea are not fastened with iron nails but with beads from the palm tree fiber which are then held with soft wood that has been soaked in water. People in Yemen fasten their ships with palm fronds. As for the magnetite mountain, it is situated on the coast of the Indian ocean; huge stones similar to the ones used in buildings were cut and used to build a square house, the top and floor of which are also made of this stone. They made a buddha, which is an idol they worship, of iron, and placed it in the middle of this house in the air, where it remained suspended due to the strong attraction of the magnetite and the same level of attraction from all sides.

Good and bad qualities

The best magnetite is the one that strongly attracts iron, is of blue color, thick, and not too heavy.

Characteristics of the stone

Aristo mentioned that if magnetite is soaked in garlic and onion water and left for a period of three days, it loses its property of attracting iron; and if one wishes this property to return, the stone should be soaked for three days in blood of a young mule, the blood being changed every day. The stone will recover its property. Another person said that if this stone that attracts iron is rubbed with garlic, it loses its magnetic property, but if placed in vinegar, it recovers it.

The judge Abul Fath Ahmad ibn Mutraf[1] said in his *Book on the Seven Stones,* which he called *Classification [of Stones],* that he mentioned magnetite, and I found this written with his own handwriting in this book. He said: I placed a stone of this kind in a perfume box containing a little musk, ambergris, camphor, and incense[2]; the stone lost its power of attraction, after being able to strongly attract iron. I then treated it as it was mentioned, washing it with vinegar, but it still did not attract anything. He said: It is still in my possession without any powers.

Properties and benefits

One property reported by Ahmad ibn Abi Khalid known as Ibn al Djazzar in his *Book on Stones,* after Salmawayh[3] the physician, is that magnetite is beneficial for gout in the hands and feet if it is held in the hand; it is also good against tetanus. Another property mentioned by Aristo is that if a woman holds a magnetite stone in her hand,[4] it facilitates her giving birth. Another property is that if a person drank iron powder or poison found sometimes in iron, or if he was wounded by poisonous iron, and this stone is powdered, dissolved in milk or lukewarm water and given to the person poisoned by iron, it will bring the poison out of his stomach by inducing vomiting until nothing of it is left whatsoever. The poison will be counteracted and the person saved. Also, if it is powdered and sprinkled over a wound caused by poisonous iron, the inherent property of the stone will cure the person on the spot. Another property is that if it is powdered and mixed with milk from a slave girl, it cures wounds caused by spears and swords. Another property is that if it is held in the hand or hung on arthritic people suffering from any kind of arthritis, they will be cured.

Aristo mentioned that amongst magnetic stones, there are those that attract[5] gold, silver, brass, [unclear word], lead, hair, and nails. He said: The stone that attracts gold is a yellow, slightly dusty stone, hot and dry. If gold is powdered with iron files and mixed with sand, and this stone is passed over it, it picks up the gold from the sand until the last bit of it. As to the stone that attracts silver, it is a cold and moist stone, of white slightly dusty color. If a person works with it, it creaks like lead, although it does not contain any trace of lead. No other stone can attract like this stone due to its inherent power. If an ounce or less of this stone is taken and placed at a distance of 5 cubits away from silver, it attracts this silver even if it has been fastened with a nail and removes the nail from its place. As to the stone that attracts flesh, there is an animal variety and another one nonanimal. As to the animal variety, it is the head of the seahare. Aristo had mentioned that the seahare has on its[6] head a stone that sticks to meat wherever it finds it if this meat is hair-free; it remains stuck on it until it uproots it, leaving a bad abcess that does not heal. Blood does not come out from the place where the flesh has been pulled out. As to the other stone, it is of a nonanimal nature, and if it gets stuck in the flesh of a living animal, it uproots it; but if it gets stuck to some dead flesh, some of this flesh peels away from the stone. If it is cooked, it becomes like crushed meat. These stones that attract as per our description of what Aristo mentioned, saying: If some of them are calcined, any one of them, in furnaces lit by fires, furnace after furnace, and sulfur stones are added on them, they burn anything that comes near, like fire does.

Value of magnetite and price

One ounce of the genuine strong magnetite is valued at 1/4 dinar.

Notes

1. Unidentified person.
2. *Nadd* is incense of aloe-wood, with ambergris, musk and frankincense.
3. Salmawayh ibn Banan, physician of the ‘Abbasid caliph al Mu‘tasem (ninth century A.D.)
4. Literally: in her bosom.Therefore the sentence should be as above, or “if a woman hangs a magnetite stone over her bosom.”
5. Word uncler in this version. “Attract” is taken from the Cairo one.
6. Literally: has a head of stone.

Chapter 15: On Emery

CAUSE of formation of the stone in its host rock

The formation of emery is the same as the one mentioned earlier under the formation of diamond, except that it is much inferior to it and is inferior in its characteristics, as if it were an inferior variety of diamond.

On the mine [location] where emery is formed

It is according to what we have mentioned earlier; it is said that it is found with diamond, and it is mentioned that the valley where emery is found is in the farthest end of China, in an island in the sea that no one reached before Alexander.

Good and bad qualities

Emery looks like crystallized sand; it comprises large and small crystallized stones, the best being the large clear ones.

Properties and benefits

Emery is cold in the second degree and hard in the third. One of its properties is that if it is powdered, it becomes more effective than when it is whole, scratching all the stones if rubbed against them either when dry or when moistened with water. It can give a strong polish and clarity to the teeth; it is slightly astringent and is used in adsorbent medicines and in medicines that fight flabbiness of the body and changes in the teeth. If it is burnt with fire, then powdered and placed on chronic ulcers and pustules in the eyes, it clears them. It is useful in adsorbent and cauterizing ointment preparations and for flabbiness of the gums.

Value and price

The ounce of emery is valued at 1 *Dirhem* or so.

Chapter 16: On Malachite

CAUSE of formation of the stone in its host rock

Aristo said that when copper solidifies in its mine, vapors similar to verdigris resulting from the sulfur produced by it arise. When they reach some cavity within the earth, they condense and solidify into a stone, such as malachite. Pliny said that malachite, lapis lazuli, *shazinah,*[1] and all the coppery stones started by being copper in their host rock. When mercury started in its host rock and was mixed with sulfur, heat overcame the moisture present in the host rock, which was in the process of becoming mercury. When the heat increased, it became red like *shazinah* and all other red stones, and if the host rock still contained some moisture, it solidified as a green stone as it absorbed some thick copper substances, such as malachite. If drought in the earth increases, its black color increases, and, together with the heated green shade in it, becomes blue, resulting in lapis lazuli and other brittle coppery stones, according to the increase or decrease in heat, cold, moisture, and drought. The person who knows how to deal with these stones can extract from them copper smoother than that found in the copper mines. Ya'qub ibn Ishaq al Kindi mentioned that if malachite is crushed with natron[2] and oil, a good, smooth, fine red copper is obtained.

On the mine [location] where it is formed

Malachite is only found in copper mines, the reason for that being what we have mentioned about its original formation from heat, but it is not found in each one of the copper mines. It is mostly found in the mines of Kirman and Sijistan in Persia and some is brought from Ghar Bani Salim in Barriyet al 'Arab. In general, it is found in many different locations, depending on the copper mines, but the best varieties amount to four: *Afrandi* [foreign], *hindi* [Indian], *kirmani* [from Kirman, Persia], and *karaki* [from Karak, Jordan].

Good and bad qualities

The best malachite is a deep green one similar in color to emerald, which is well known for its beautiful color, showing crescents and nice eyelike formations, hard and smooth, readily accepting a good polish. These are the qualities of the genuine one, found exclusively in the *afrandi* variety.

Characteristics of the stone

Malachite is a soft stone; one of its characteristics is that if table utensils or knife hilts are made from it, after several years they will become scratched due to its softness, and they will lose their brightness. Another characteristic is that it is easily dissolved, and it can be cut into beads or utensils, or other objects. Ya'qub ibn Ishaq al Kindi mentioned in his *Book on Stones* that he had seen a malachite saucer that could hold 39 *Ratls.* Another characteristic is that if it is soaked in oil, its green color deepens, and its color improves. But if it is left too long in oil, it acquires a blackish shade. Another characteristic is that if a person drinks its powder, it is poisonous; it tears the intestines and causes the body permanent damage. Another characteristic is that if a person drinks some of its powder or holds it in his mouth and sucks it, it causes him harm. The person should be treated by drinking aged wine and including butter and grease in his food; he should be treated as if he had drunk verdigris.

Properties and benefits

One property is that if it is rubbed over a place bitten by a scorpion, it eases some of the pain. Another one is that it prevents pregnancy when taken orally, but this is dangerous, as we have mentioned before. Another property is that if some of it is powdered, added to vinegar, and massaged into pustules caused by the black bile, it clears them. It is also beneficial for head colds and general ones. One property of the *Afrandi* variety that has been tested is that if it is dissolved in water and used as an eye drop, it is good for whites in the eyes, which it removes after three applications. Another property is that if its powder is drunk by a man who has drunk poison, it will be of some benefit for him. Another property is that if a person who did not drink poison takes it orally, it tears his intestines, burns the body, and kills rapidly, especially if it was abraded with iron; then it destroys the body, bursts the intestines, and is deadly. Another property is that if a small amount is powdered and added to gold that has broken under the blows of the jewelers due to its roughness, it straightens it, smoothes it, removes its roughness, and improves it. If it is mixed with tinkal, it makes it more effective. It is also said that it gives [a red color to gold and colors it];[3] it is moderate in heat and hardness; and it is said that it is hot in the fourth degree. Another property is that if it is used to kill seven flies, then is crushed and dissolved in water and spread over scorpion bites, it cures them.

Value and price

The pure, genuine *Afrandi* variety with all the above-mentioned qualifications is worth 2 *Mithqals* per *Mithqal.*

Notes

1. Unidentified red stone.
2. Natron is a native salt of Egypt.
3. This sentence is missing from this version. It was taken from the Cairo one.

Chapter 17: On Lapis Lazuli

CAUSE of formation of the stone in its host rock

This was already mentioned under malachite.

On the mine [location] where it is formed

Lapis lazuli is brought from a mountain in Tukharistan from a place called Jestan in the land of Persia near Armenia.

Good and bad qualities

Lapis is a soft, porous stone. The best is the most radiant and clearest, with an even sky-blue to midnight-blue color.

Characteristics of the stone

One characteristic is that if it is joined to gold, each one increases in beauty in the eyes of the onlooker, although they do not alter in nature, and their individual characteristics do not improve or become worse, except that the color of each one improves the other in the eyes of the onlooker, as if they complemented each other.[1] Another characteristic is that if a piece of lapis lazuli is placed over smokeless coal, the resulting flame will be colored blue; while the lapis lazuli stone remains unchanged. Genuine stones are differentiated from the fake ones with this test.[2] Another characteristic is that if it is calcined, it calcinates and retains the fire within it.

Properties and benefits

One benefit is that it is good for the eyes if it is added to collyriums. Another benefit is that it promotes growth of the eyelashes and clears moisture in the eyes. It rejuvenates the roots of the eyelashes, increasing their number, making them grow and strengthening them. It must be well powdered and used as [other] powders are used. Another benefit is that, if it is drunk after being washed,[3] it provokes diarrhea, and if drunk unwashed, it induces vomiting. It also safeguards against melancholy and clears away the black bile. If 4 carats [of lapis lazuli] are drunk with rose syrup diluted in lukewarm water, it is of outstanding benefit for quartern fevers, [as it reduces substantially the gastric juices resulting from the black bile like no other medicine can].[4] It is also good for asthma; it can increase menses, is good for pains in the bladder, and can remove boils. If it is powdered on a piece of wood from the palm tree, it is useful for open ulcers that eat at the skin, and if powdered with vinegar and spread over leprosy spots, it fades them away. If drunk after being washed, it can extract the black bile without causing any fever in the body. And if drunk without being washed, it extracts it by inducing vomiting. Moistened with honey, it is useful for pains in the liver. Also if powdered [and mixed] with vinegar, then spread on leprosy spots, it is of benefit. Also, if it is hung on a young boy, he will not afraid. Also, if added to a hair cream–any cream–it makes the hair curly and improves its texture. It is a cold and hard[5] stone.

Al Maseehy[6] said in his book entitled *Al Mi'at Hadjar* (The Hundred Stones) that lapis lazuli is effective against the plague without causing any harm or disease. But this is not easily done, and it is slow-acting; it should be washed several times with pure water and should be well powdered; otherwise it can cause vomiting and nausea; [then] it is mixed with some aromatic spices and the average dose [of this medicine] is 2 1/2 *Mithqals.*[7] The Armenian stone[8] has the same action and is washed in the same manner and is used for the same remedies, but its average dose is 1/3 *Mithqal.*

Lapis lazuli can be manufactured according to the manner that I shall describe specifically: A portion of red *zarnich*[9] is taken, and another equal portion of yellow *zarnich*[10]; also a quarter portion of vitriol from Kirman and a similar [quarter portion] one of clear quartz taken from pure sand. Each [ingredient] is powdered separately, then sieved and moistened with vinegar. A clay container is taken and coated well with compact clay containing hair and manure, then it is left to dry. The ingredients are placed in it, moistened with just enough vinegar to reach the consistency of *sawiq,*[11] then the clay container is closed with rags, and its aperture is coated with clay. The oven is lit with wood then with manure until the manure in it becomes as [thick as] the bone of the arm, and the clay container is placed in the oven, buried under the manure. The oven is closed with clay, lit at the bottom, and left overnight. In the morning, the clay container is removed from the oven, and what is inside is taken out. You will find small red stones, similar to ruby, round like pearls or elongated, the best one can see of stones. This is how mineral lapis lazuli is made. Nothing was omitted. However, I have mentioned the above to let you know that there is a manufactured mineral variety of lapis lazuli. It is the easiest thing to fake and to imitate and can be made according to several methods other than the above-mentioned one, but this one is the best as it is also used in engravings and dyes, and it is not more compact than the mineral one.

The way to test the genuine mineral lapis lazuli wanted for use in medicines is with the flame test,[12] as we have shown earlier. If it remains stable and does not lose its color, then it is genuine.

Value and prices

Genuine lapis lazuli is found as a large stone or a small one; the large stone is panned, purified, and ground, others are [used for] finger-rings. The good genuine stone suitable for a finger-ring [is worth] approximately 3 *Naqara Dirhems*, and the genuine stone that has been panned, purified, and ground is worth 1 dinar per ounce or less than that. The rough stone is worth 2/3 of the price of the purified one or more, but it loses approximately 1/3 of its original weight during the panning and purifying process.

Purifying and panning lapis lazuli is a profitable business,[13] exercised by those who specialize in this work and live from it. This is done as follows: The genuine [pure] mineral stone tested over the fire according to what we have already mentioned before is taken. It is made into a dough[14] consisting of a portion of pure resin to which is added a portion of pure frankincense. This dough is placed in a brass smelting pan over low fire until it melts. The lapis lazuli is then powdered, kneaded with water, and placed in the pan; the ingredients are stirred with a solid brass rod until they are well mixed. Then they are covered with pure water. At that stage, the mixture solidifies, so the fire is gradually increased until it melts again, while continuing to stir with the above-mentioned rod. If the essence[15] of the stone comes out, it is the pure genuine lapis lazuli that was easily obtained. But if the purified lapis lazuli does not come out by this operation, some substances are added that extract it. This is the secret of the profession that only few people know; [as a matter of fact] this knowledge is withheld by the people who manufacture it. For the lapis lazuli can be ruined at this stage if the secret [of its manufacture] is not known.[16] I have not taken this information from any book, but I have acquired this knowledge through experimentation according to my[17] books on the synthesis of gemstones. If it does not come out, the essence of the lapis lazuli can be extracted with olive oil or with soap made from olive oil. Whichever is available is poured [on the mixture], then the lapis lazuli will cast its color, and its essence will come out[18] until none of it is left in the remaining background at all. It is then poured in a clean well-oiled porcelain or clay bowl and left until all the residues and other impurities of the stone mixed with its essence sink to the bottom, and the lapis lazuli pigment and purified essence that float on top are taken. This operation causes the stone to lose 1/3 [of its original weight] more or less, depending on the quality of the stone, the correctness of the procedure in extracting its essence as I have explained; also, ignorance and errors can destroy most of the stone or all of it.

Notes

1. Al Tifaschi was probably referring to pyrite inclusions in lapis lazuli.
2. This test is called the element flame test in modern chemistry.
3. To wash lapis lazuli or to purify it is to extract the ultramarine pigment.
4. This sentence is missing in this version. It was taken from the Cairo version to complete the meaning.
5. Literally: dry.
6. Abu Sahl al Maseehy (died c. A.D. 1000), physician, teacher of Avicenna, wrote a 100-chapter treatise on medicine (*Al Kutub al Mi'at fi al Sina'at al Tibbiyya*). This could be the one referred to by al Tifaschi.
7. 2 *Mithqals* in the Cairo version.
8. The Armenian stone used to describe lapis lazuli in ancient and medieval times. Al Tifaschi seems to differentiate between the Armenian stone and lapis lazuli in his book. Therefore, by "Armenian stone," he could have meant sodalite.
9. Red zarnich is red arsenic or realgar.
10. Yellow zarnich is yellow arsenic or orpiment.
11. *Sawiq* is a mush made with sugar and dates.
12. Literally: placing on the coal. Al Tifaschi refers here to the element flame test.
13. Literally: industry, profession.
14. The word "dough" in Arabic could also mean nucleus, basis, or yeast preparation. In this context, the word "dough" is the most appropriate.
15. Meaning the ultramarine pigment.
16. The sentence between brackets could also be interpreted this way: "This is where the craftsmen can ruin the stone."
17. The word "my" could also be interpreted here as "our," meaning other books.

18. This means that the ultramarine pigment will float on top of the surface.

Chapter 18: On Coral

CAUSE of formation of the stone in its host rock

The formation of coral is midway between the two states of inorganic substances and plants; it resembles minerals in petrification and resembles plants in its tree-like formations that grow from the bottom of the sea, with stems and erect green ramifying branches, [although] not fruit-bearing.

Pliny said that the cause for this is the mixing of heat and moisture in the core of the mineral, and the preponderance of the moisture over the heat due to its proximity to the water. Coral resembles a mineral in its body and resembles a plant in its soul; as when water remained a long period of time on the land and became heated under the heat of the sun, it was able to dissolve the dryness of the land with its inherent moderation, which it acquired from the heat of the sun. When its various parts dissolved due to mildness of the water and became hot and temperate, and as the sun heated it with its strong heat, it was able to get rid of the heat and drought of the sun on account of the dry, hot substances in its essence. And when the dry parts in it condensed together, and the heat reached it while still moist, it came out into the open, and upon contact with the air, it lost the heat inherent in it and cooled and solidified, becoming a hard stone. The inherent characteristics [of the stone] were still effective; thus, whenever this water avoided the heat of the fire, it exerted pressure on what followed and rose in the air, and the heat of the fire affected it, and it developed into a high plant of white color. It turned red due to its loss of moisture and dryness. And when the heat of the sun increased, it removed all moisture from its surface, making it red on account of the strong heat. This is the cause of its red color. I said that it was a plant because it grows in the air and ramifies into branches like plants. It developed branches as when the water mixed with its dryness and became heated with the fire, it took refuge from it in the air where it developed branches in accordance with the amount and strength of heat present. This is the cause of the ramification of its branches, and this is the cause of its formation in its host rock.

Ahmad said: Likewise, some animals are formed in the water that belong midway between the animal and the mineral worlds, such as the aquatic sponge, which grows on the stone and remains fixed [in its place]; if it is touched, and it feels this touch, it cringes, but it does not have the locomotive ability of the animals.

In the coral sea, divers take strong nets made of rope, weighted with lead; they wrap them around the [coral] tree and pull the net until it is uprooted. It comes out from the water soft and of white color, and when the air hits it, it dries up and becomes red. Its roots, called *Basaz,* are removed, and [then] its branches, or coral, are sorted out into large and small pieces, according to the knots and ramifications present in it. They are ground with water on the whetstone and polished with emery. Then, if wanted, they are drilled with tempered steel.

On the mine [location] where it is formed

Coral is found in a place called Marsa al Kharaz in the sea of Africa.[1] It is also found in the sea of Europe,[2] but most of it comes from Marsa al Kharaz, from where it is brought to the East, Yemen, India, and all countries. In no other place are larger, bigger, or better quantities found.

Good and bad qualities

The best is the one with the largest size, undistorted, of deep red color, free from wormwood-like cavities, which are holes found inside it, some of them hollow like bones, which make it defective. Also, the ones with a whitish shade and lighter red color are defective. Knots and scratches also constitute defects, but they are virtually always present, as coral grows in ramifying branches, as we have mentioned. Rarely, a solid undistorted piece of coral is found, free of knots and scratches, and if it is found, and its color is deep red, then it is of top quality. If a large scratched piece is found, it is scraped until its scratches and knots are removed, and it becomes smooth and of regular shape, but this work takes away a lot of its weight. The increase in price depends on its quality. Large and rare pieces of coral are removed for the king of Africa, for whom they are made into inkstands and knife hilts. I have seen such an inkstand, one and the half span of the hand in length and three fingers both in width and in height, of the best red color, purity, and quality.

Characteristics of the stone

One characteristic is that if it is soaked in vinegar, it softens and becomes white, and if left in it [for a long time], it dissolves. Some people use it in finger-rings, and if a person wants to write something he likes [on coral], he covers the whole stone or the whole ring with wax; then he engraves with the tip of a needle what he likes until the wax is removed from the place of the engraving only. Then he soaks the stone in strong vinegar for one day and one night, or two days and two nights, then he removes it and removes the wax from it; he finds the place where the etching is engraved, eaten away by the vinegar, while the rest of the stone remained as it was, unchanged. I have tried this and obtained what I have mentioned above.

Another characteristic is that if it is soaked in oil, its red color brightens and improves, and it has the opposite effect of vinegar. Coral is cold in the first degree and hard[3] in the third; it contracts and cools in a moderate manner.

Properties and benefits

One [benefit] is what Alexander mentioned, namely, that if coral is hung on an epileptic person, it is beneficial for him. Another benefit is that it safeguards small children from the evil eye and the evil spirits, human or occult, if it is hung on them. Another benefit is that if it is burnt and rubbed on the teeth, they become white and well polished; it removes their cavities and strengthens the gums. The way to burn it is to take some stones, put them in a new clay container, then close the lid with clay. It is then placed in an oven that has been lit throughout the night.[4] The following day, the stones are removed, powdered, and used. It is also beneficial for eye ailments; it can dry out moisture if used in a collyrium and can remove excess meat from ulcers and fade away their traces. It can also fill up old ulcers and is beneficial for darkness of the eyes as well as whites and oozing of the eyes. It is also beneficial for heart palpitations if drunk in powder form and is used in medicines that remove blood clots and are used for weakness of the heart. Also if drunk, it is of substantial benefit for blood hemorrhages and abrasions. It is also beneficial for dysurea when it is drunk. Also, if drunk with water, it removes the swelling of the spleen. Also, if hung over the stomach, it is of enormous benefit against all types of stomach disorders; it is even said that it is of similar efficiency in this respect to the *zhubabi* emerald. It also stops hemorrhages from any part of the body.

The way to use it as an oral preparation for bleeding and blood discharges is to powder it after burning it as described earlier, then take 3 *daneqs* of this powder, add to it 1 1/2 *daneq* of gum Arabic and knead it with egg white. [The mixture] is drunk with cold water and is of great benefit for hemorrhages and blood discharges. Another [benefit] is what Alexander said, namely, that coral is beneficial for gout in the leg if hung on it.

Value and prices

In Africa[5] where coral is found, as we have explained earlier, the value of 1 Moroccan *Ratl* is from 5 to 7 Moroccan *Sakkiyya* dinars, each dinar being worth 10 *Sakki Dirhems* in their language, which is equivalent to 5 *Naqara Dirhems.* This is the price of the *"daq,"* which is the rough variety, not engraved or worked. It is polished, engraved, and worked in Alexandria, thus its value in Alexandria is double that in Africa which we have mentioned, and may be three times the price, depending on supply and demand. Large pieces are more sought after and command higher prices than the small ones.

Notes

1. Africa refers to Tunisia here.
2. This refers to the Mediterranean waters around France and Italy.
3. Literally: dry.
4. Or: placed in an oven which is then lit throughout the night.
5. Meaning North Africa here.

Chapter 19: On Jet

CAUSE of formation of the stone in its host rock

Jet is one of the lead stones, and we have already mentioned that every stone that can be referred to one of the base stones started initially by being one of them, but external factors such as increase or decrease in heat, cold, moisture, or drought prevented this, as we have explained earlier.

On the mine [location] where it is found

Jet is brought from two places, one being India, the other Persia.

Good and bad qualities

The best is the Indian one, which is a deep black stone without transparency except that one's face can be seen in it like in a mirror. It is shiny, very brittle, and [can be] quickly broken.

Properties and benefits

Jet is cold and hard in the first degree. It is useful when added to collyriums. Aristo said that one property of jet is that if a person's eyesight has been weakened due to old age for some reason, thus making him see things through a shadow similar to a veil or flies or fog; all this is an indication of the beginning of cataract in the eye; [then this person] takes a piece of jet for a mirror and stares at it, his eyesight will be strengthened and it will cure its ailment. Another benefit is that if it is worn in a finger-ring and stared at, it sharpens the eyesight. Another benefit is that if a person wears a jet bead, it protects him from people wearing white clothes.[1] Ibn Saharbakht said that it is good for the eye if it is powdered and used in collyriums; as it sharpens the vision. Also, if it is mixed with seeds of *kulthum* [meaning unclear] and the gallbladder of a sparrow hawk and musk and is used in a collyrium, it has the same effect.

Value and price

A bead of 1 *Mithqal* is worth 1/2 *Naqara (Dirhem).*

Notes

1. A person of white color in the Cairo version.

Chapter 20: On Amethyst

CAUSE of formation of the stone in its host rock

Amethyst is an iron stone[1] that started in its host rock by being iron but was prevented by external factors, namely, increase or decrease in the four elements.

On the mine [location] where is is found

Amethyst is found in a village called al Safra at a walking distance of three days from Tayba the good city[2] of the Prophet of God, peace be unto Him.

Good and bad qualities

There are four varieties of amethyst. The first one is the best one and is of strong pink and blue colors and is the most valuable kind, followed by one of strong pink but weak blue colors, followed by one of strong blue and weak pink colors, followed by the cheapest and lowest kind, of both weak blue and pink colors. Amethyst is a stone that the Arabs appreciated and used for decorating their utensils and weapons. The way to cut it and polish it is the same as that for emerald; I mean that it is first ground with emery on a bed of lead with water; then it is polished according to the purpose wanted. One property is that which is mentioned by the Arabs, namely, that it encourages its wearer in times of battle, also what al Razi mentioned in his book, which he called *Tuhfat al Muluk* [The Gem of Kings], namely, that a characteristic of amethyst is that whoever makes a drinking glass of amethyst and drinks in it as much wine as he wishes, he will not get drunk. Another property is that it protects its wearer from gout. Another property is that if it is placed under a pillow, it protects from nightmares.

Value and price

A stone weighing 1/2 *Dirhem* is worth 1 *Dirhem.*

Notes

1. In other words, iron enters in the composition of this stone.

2. Medina, Saudi Arabia.

Chapter 21: On Hematite which is [the stone] called *Hadjar al Sirf*

CAUSE of formation of the stone in its host rock

This stone is a black iron-based one; the formation of iron stones has already been discussed.

On the mine [location] where it is formed

It is brought from Karak[1] at a walking distance of seven days from Egypt, from where it is carried to other countries.

Good and bad qualities

The best is the deep black one with an iron red shade.

Properties and benefits

This stone is beneficial against the ill effects of drinking pure unadulterated alcohol. It is known for this [property] and was named accordingly. This is done by filing it and drinking its powder. Its effect and benefits are apparent.

Value and price

1 *Ratl* of this stone is worth 3 *Naqara Dirhems* in Egypt. But it is more expensive in countries other than Egypt, due to the closeness of its mine to Egypt.

Notes

1. Karak, Jordan.

Chapter 22: On Nephrite

CAUSE of formation of the stone in its host rock

Nephrite and jasper–also called *Yassb*–are two silver-based stones of similar composition; they are formed in silver mines from vapors still not condensed into silver, following an increase or decrease of the four base elements, according to what we have explained earlier.

On the mine [location] where it is formed

Kashgar from where it is brought to all other countries; Kashgar being a province in it with a large town [situated] between China and the town of Ghaznah, at a distance of twenty odd days from Ghaznah to the north, their language being Turkish.

Good and bad qualities

Nephrite comes in two varieties: a white one and one of yellow color like that of old ivory. This is the good, genuine mineral variety. As for the white variety, it is manufactured in China from a mixture of substances and does not possess any of the properties of nephrite except that it looks like it.

I have myself manufactured in Cairo–may God protect her–some utensils of this nephrite, which I gave to an emir who buys, collects, and values nephrite items. He did not suspect that what I gave him was not manufactured in China, and when I informed him that I was the one who manufactured them, he refused to believe me until I presented him with the proof thereof and made for him utensils of specific shapes; then he believed me.

Properties and benefits

One characteristic of this stone is that lightning never hits a place where this stone is present. Some trustworthy persons from Persia told me that they had seen towers in Persian lands where lightning often hits. Thus, lighthouses were built on these towers [or forts] where this stone [nephrite] was hung well in sight. One can see lightning coming down from the sky avoiding the place where the stone is [hung], but hitting all other places far from it. Another property is that it decreases the recurrence of erotic dreams. Another benefit mentioned by Galen in his *Al Adwiya al Mufrada* (The Simple Medicines) is that it is beneficial for pain in the stomach if hung on it externally.

Value and prices

Nephrite is made into various bowls in China. I have seen such as bowl that was sold in the market in Cairo for 5 Egyptian dinars. I have also seen another large bowl for which its owner asked 50 dinars. As for the nephrite finger-ring, it is worth 5 *Naqara Dirhems*. Some [stones] are manufactured in India and sold as nephrite, while in fact they are not.

Chapter 23: On Jasper

CAUSE of formation of the stone in its host rock

We have talked about this earlier.

On the mine [location] where it is found

In Yemen, from where it is brought to other countries.

Good and bad qualities

Jasper comes in white and olive green, the olive green one being the best quality. There is a blue variety, but it is dyed.

Characteristics and benefits of the stone

Al Maseehy said that jasper is specifically beneficial for all kinds of stomach ailments if hung [on it] externally. Ahmad said: This is what was reported after al Maseehy. But I believe that this was an incorrect report and what was meant was the aforementioned nephrite and not jasper –God knows best. A characteristic of this stone is that it can [easily] be dyed, and jewelers dye it to imitate emerald. It is mostly found as small stones.

Value and price

A stone weighing 1 1/2 *Dirhems* is worth 1 *Naqara Dirhem.*

Chapter 24: On Rock Crystal

CAUSE of formation of the stone in its host rock

Pliny said that rock crystal is a borax-based stone, its color being white due to the factors that affected it. It originated as a ruby [or corundum] similar to silver, which is basically a borax stone and originated as gold; likewise rock crystal, I say that the host rock where rock crystal is found contained moisture mixed [with dryness], and when it was heated, the moisture overcame the drought, and when the heat of the sun reached it, it was heated, and it permeated the dry solid [substance], dissolving it by its gentle action over a long period of time. When it was dissolved, the dry part in the moisture became pure water as it was overcome by the moisture and had matured under moderate conditions. And when it dried, it solidified and became a pure white stone. The moisture [in the place] and the moderate conditions in its host rock prevented it from becoming a red stone; instead, it became a stone of outer white color and inner red core.

Rock crystal disintegrates in the fire due to its salt content, which sprang forth from the little grease on its exterior [mixed] with the cold apparent on its surface; thus salt appeared, and the grease remained inside the stone with the heat; thus the grease was intermingled with the salt, and when touched by the heat of the fire, this salt disintegrated, causing the whole solid also to disintegrate. Iron affected it due to its high content in moisture, which tempered its drynes, thus it became soft and brittle, but clear [limpid] due to the little compactness of its parts, which did not compress and sinter due to the little dryness in it and little effect of the heat in its host mine. This is the cause of formation of rock crystal.

On the mine [location] where it is formed

Some rock crystal is formed in Djazirat al 'Arab in Hidjaz, this being the best variety. Another variety is brought from China, which is inferior in quality to the Arabian one. Rock crystal also comes from European countries, which is also of good quality. There are also mines near Armenia, which yield rock crystal of yellowish tinge, which is known as the "glassy" type, as if it were cooked in the fire. Nowadays, new mines were discovered in the furthermost *Maghreb*, near Marrakesh, the capital of Morocco, yielding clear but cracked rock crystal. It became so abundant there that the king of Morocco used rock crystal for decorating the walls of a large *Madjlis.*

Good and bad qualities

The best rock crystal is the clearest, most limpid and transparent, of strongest white color, free from cracks. If, in addition to these qualities, it is of large size, whether made into utensils or other items, it is the top quality rock crystal. Ya'qub ibn Ishaq al Kindi said that some rock crystal pieces yield stones of over 100 *Mna.*

The author of this book said: One of the European merchants gave the king of Morocco a dome made of two pieces of rock crystal, able to seat four persons. In Europe, curious containers are made from it. I have seen in the possession of a king of Africa rock crystal carved into the shape of a rooster, given to him by some European person, able to contain 4 *Ratls* of liquid, which penetrates everywhere in the rooster, even in its claws, as it was all hollowed out. I have seen its claws; if liquor is poured in it, it reaches until these claws. Some dirt accumulated in the neck of this container, blocking it, and a person was requested to clean it, which he could not do due to the danger involved in removing the dirt. A cutter was sent for, and he asked 50 Moroccan dinars to complete this task, but the king was kind and generous to him until he agreed. He took the container and cleaned its neck without letting anybody know how he did it, making it like new. I befriended the jeweler until he told me how he did it; he showed me a male lizard that he had used for this purpose, as the male lizard is well known to squirm in humid surroundings. He had introduced it with a lathe in the neck of the container, turning it inside with the lathe until it removed the accumulated dirt in the process.

Many persons from Persia told me that close to Ghaznah, between it and Kashgar, at a walking distance of thirteen days from Ghaznah, there is a valley between two mountains, where there is a road leading to Kashgar. These two mountains that are on each side of the valley are mountains of pure rock crystal, which are cut out at night because the reflections from the stones, when hit by the sun, prevent working them during daytime. Water containers are made from rock crystal in Kashgar and in Ghaznah. A person from Ghaznah told me that he had seen in the castle of the king of Ghaznah, Shihab al Din al Ghaznawi, four water containers, each able to hold three water bags, of the kind carried on mules. These were placed on supports from which the containers were reached by a staircase; the containers and the supports being all of rock crystal.

Characteristics of the stone

One characteristic is the one that Theophrastus mentioned in his *Book on Stones,* namely that rock crystal melts like glass and can easily be dyed. Ahmed said: This is true, but this is not due to an inherent softness in the rock crystal, but as a result of external factors; otherwise if rock crystal is placed in the fire, it disintegrates and calcinates.

He said: Some people who visited Kashgar told me that their water containers were made of rock crystal, obtained from the above-mentioned nearby mountain. Another characteristic is that if the sun rays are captured within it, and a soft rag is positioned where one sees the refraction of the light coming out of the stone, the rag burns, and fire can be obtained from it.[1]

Properties and benefits

If it is worn by a person, he will not see any nightmares.

Value and price

The value of rock crystal differs according to the size of the stone and the utensils made from it. The pleasure in it is derived from the large size of the utensils or other items made from it. In general, the piece that can hold 1 *Ratl,* if it is perfectly clear and free from cracks, is worth 3 Egyptian dinars or so.

Notes

1. This is a reference to the use of rock crystal as magnifiers or for cauterization of wounds through the agency of the sun. In the West, the theoretical discovery of magnifying lenses was made in Italy in the fourteeenth century. However, Ibn al Haytham (A.D. 965-1039), a famous physicist and student of optics at the court of the Fatimid sultan al Hakim of Egypt, had presented experiments for testing the angles of incidence and refraction which probably laid the foundations for this discovery. (Hitti, *History of the Arabs*, 629).

Chapter 25: On Talc

CAUSE of formation of the stone in its host rock[1]

Talc falls from the air like dew, solidifying when it reaches the ground, one layer over the other. It formed when the thick humidity was overcome by land encroachment and drought, causing its parts to solidify and sinter together, without any presence of inherent greasiness, and it was not overcome by drought, as it originated from water; for this reason, it does not melt in the fire like the base elements.

On the mine [location] where it is formed

Talc is found abundantly in the island of Cyprus, from where it is brought. It also comes from many other places, and I was told that there is a mineral variety of talc.

Good and bad qualities

Talc comes in two varieties: silvery and golden. The silvery is of pure white color, the golden has a tinge of yellow, and is the best variety. Talc is cold and hard.[2]

Characteristics of the stone

One characteristic is that if it is placed in fire, it does not burn or calcinate or melt like other stones. For this reason, the sages say that if it is dissolved and coated on bodies, it will shield them from fire. Another characteristic is that if it is struck on a brass anvil with an iron hammer or anything used for striking solids, it has no effect on it. The only trick that can be used for powdering it is to take the portion required, wipe it [?] and place it in a black hair cloth or in a coarse piece of cloth with [some] small stones; then place the cloth in hot water where fava beans have been cooked. The talc will gradually dissolve completely, then it is gathered like pounded flour, and you can use it as you please.

Properties and benefits

It disintegrates stones in the kidneys and is beneficial for the bladder if its powder, extracted according to the manner shown above, is drunk. Another property is the one mentioned by al Maseehy [unclear word) in his book entitled *Al Mi'at fi al Sina'at al Tibbiyya*[3] [The 100-Chapter Treatise on Medicine], that talc is used externally to stop hemorrhages, and that it should never be used internally. Another characteristic is what Ibn Saharbakht said, namely, that it is useful for malignant growths of the breasts, the testicles, and the joints between the thighs and the abdomen at their onset, also for hemorrhages in the uterus and posterior, as well as for ulcers in the intestines.[4] It is also useful for blood discharges if it is dissolved in water from the plantain[5] and can counteract poisons if it is drunk with wine.[6] He said: It is cold in the first degree and hard[7] in the second [degree]. And we have already mentioned that al Maseehy said it should only be used by hanging it externally. With regards to what the ancients said about using talc for shielding bodies from fire, I have reported the [following] based on al Mas'udi's book *Asrar al Khulafa* [The Secrets of the Caliphs] when stating the recipe for coatings that are used on weapons and horses to make them fire resistant: 1 *Ratl* of talc and 4 *Ratls* of gum Arabic are taken; also 2 *Ratls* of gypsum; and the required quantity of chalk flour, and for each 10 *Ratls* of all the above ingredients, [add] one *Ratl* of pulses.[8] The talc is powdered according to the method described earlier, and it is added to the gum Arabic and mixed with the gypsum and flour. The pulses are moistened [?], water is added to the wine vinegar to break down its acidity, then it is mixed with the moistened pulses. The ingredients[9] are well kneaded into a paste that can be coated on what we have mentioned, and you can coat it on anything you like, it is good. He said: If you coat a *Djubba*[10] with it, and you place it in the fire, it will not burn. The author of this book said: Talc can be dissolved in many other ways that have been tried [successfully], but they are not used for fire-resisting purposes. I have not tried the above-mentioned method, but I have copied it as I found it [written] for it to be tested and established. Whoever describes this method will be thanked and praised by all the scientists.

This is the last thing I wanted to mention. May God, the Creator of life, be thanked and may all His Prophets be praised.

Notes

1. In the margin are written the following words: "There are three varieties of talc: *Yamani* [from Yemen], *hindi* [from India], and *andalusi* [from Spain]. The best kind is the *yamani*, the lowest variety is the *andalusi*, the *hindi* being the average one."

2. Literally: dry.
3. This title is taken from the Cairo version.
4. Peptic ulcer?
5. *Plantago major L.*
6. Literally: in a drink. I presume al Tifaschi meant alcohol.
7. Literally: dry.
8. Literally: pulses or legume seeds.
9. Literally: medicines.
10. *Djubba* is a long outer garment open in front with wide sleeves.

PART III
INDIVIDUAL GEMSTONES

CHAPTER 4

PEARL *(DJAWHAR)*

AL TIFASCHI started his book with the chapter on pearls. In old times, the pearl was called *Djawhar* in Arabic. This term referred (and it still does) to all precious stones, and according to the *Encyclopedia of Islam,*[1] the Arabic lexicographers give "jewel" as a meaning for *Djawhar* at least from the ninth century. Al Tifaschi explained: "The term *Djawhar* refers to all precious mineral stones, then it was also used to refer specifically to pearls, as it is the most valuable of gems."[2] This is proof of the high esteem with which the Arabs held this queen of gems. The term *Djawhar* was the generic name for pearls and referred to both large and small pearls, as well as to very small (seed) pearls, both drilled and undrilled. In addition, special terms were used by jewelers at the time to differentiate between pearls: Large pearls were called *Durra,* small ones *Loulou daq, Loulou habb* or *Mardjan.* (In Arabic, *Mardjan* nowadays refers to coral, and this could explain the misnomer sometimes used to describe coral beads, namely "Red Sea pearls"). Drilled pearls were called *Djumahana* or *Shazra*; undrilled ones were called *Durra, Habba,* or *Kharida.* White pearls of exceptional orient were called *Thuma* or *Khafrada.* Nowadays, pearls are simply called *Loulou* (or *Lou'lou')* in Arabic. (Note: All the above terms are in the singular form, except *Loulou,* which is the plural of *Lou'lou'a*). In North African countries, however, the word *Djawhar* still refers exclusively to pearls.

As early as A.D. 1240, the date when his book on gemstones was written, al Tifaschi knew that pearls were made of thin concentric layers, in fact, "those that are not so are not genuine."[3] The concentric structure of pearls was known even before, that is, during al Biruni's time, as he mentioned that pearls were often "peeled" to correct a defect or to prove their genuineness.[4] Al Biruni compared the structure of pearls to that of an onion, formed of thin concentric, successive layers. However, al Kindi was the first to point out the concentric growth of pearls in the ninth century, saying that pearls were formed of identical concentric layers, all made of mother-of-pearl.[5] How did al Tifaschi and his predecessors notice this structure without the help of viewing lenses or microscopes? At any rate, al Kindi's account (A.D. 801-866 or 873?) in the ninth century is the first, preceding that of Réaumur (1683-1757) by 900 years.[6]

Another important remark here is that al Tifaschi mentions the availability of imitations of pearls (probably other than glass, as glass looks very different and is very easily recognizable) as early as the thirteenth century and probably even before this period.

Pearls were found mainly in Sri Lankan waters (this island was known as *Sarandeeb* at that time), in the Gulf of 'Oman, Bahrain, the Red Sea, and the Persian Gulf. The best pearls came from Sri Lanka and the Persian Gulf; these are still the best locations for natural pearls nowadays. Pearls came in different colors: green, white, and white with a yellowish or creamy tinge. Pure white without any other hint of color was highly prized. In fact, the oysters of the Persian Gulf and Sri Lanka, known as the *Pinctada radiata* and *Pinctada vulgaris*, give creamy-white and pale cream-white pearls; while those of the Red Sea called *Pinctada margaritifera*, yield larger pearls of silver-white color.

The perfect pearl is described as being "valuable either on account of its large size and heavy weight, or on account of its qualities: strong white color, good clarity and orient, even distribution of color, completely spherical shape, and perfect general appearance."[7]

Al Tifaschi quoted the famous tenth-century Arab historian and traveler, al Mas'udi, as saying that pearl divers carry with them jars containing a type of grease that shines when immersed in water. When they see a dangerous animal, they spill some of this grease in the sea to frighten it away. These divers also used a special trick of emitting sounds similar to dog barks, which frightened dangerous sea creatures. Al Tifaschi also quoted the ninth-century Arab physician Ibn Masaweih when he described pearl divers: "The diver returns only when his eardrums explode and blood comes out. Then he keeps practicing until he is able to breathe very faintly, and when he dives he puts on a nose clip made of ivory or horn so as to keep the water out, and he breathes from this opening (within his ears) and can stay up to half an hour underwater. He effects three dives daily, from morning till midday, and his food consists of fish and dates."[8] Black stones were used as ballast to frighten sharks and other dangerous sea creatures.

Recommendations and pearl remedies

Small pearls and those light in weight, as well as difformed ones (baroque pearls) and those of yellow tinge or exhibiting large drilling holes, were considered to be of poor quality. Al Tifaschi warned that fat and grease affect pearls, as well do all types of acids, especially citric acid. He also advised keeping pearls away from sources of heat and fire, perspiration, bad odors, and hard substances, as the pearl can easily be scratched and damaged. These are the exact same recommendations found today in any modern gemological textbook.

The following are few of several remedies proposed by al Tifaschi to improve the color of pearls:

• Strong distilled wine vinegar and distilled lemon juice can clean pearls and restore their whiteness; however, it could make them gritty and harsh to the touch.

• Small yellowish pearls should be soaked in milk exhudation from the fig tree for several days, the milk changed every three days until the pearls come out white.

• One can also soak pearls in an equal mixture of soap and alkali and put them in a clay pot over low coal fire for two to three hours.

• Another method is to take some *mahlab*, peel it, and mix it with an equal amount of peeled sesame seeds and camphor. The ingredients are then powdered and moulded into a ball where the pearls to be treated are placed. The ball is placed in an iron ladle and covered with animal fat. Cooking takes place over a low coal fire until the pearls come out white.

Several other recipes, more complicated than the ones mentioned above, were also given. It is interesting to note that one of the modern techniques for treating pearls consists of bleaching them (in a manner similar to that of al Tifaschi's), whereby the pearl is soaked in some kind of bleaching agent (such as diluted hydrogen peroxide). Bleaching, however, could damage pearls.

Prices

Al Tifaschi used the *Mithqal* as a unit of weight for pearls. Mineralogists and gemologists have different opinions as to the value of the *Mithqal*:

J. Sinkankas reported on different values[9]:
5 *Mithqals* = 35 carats therefore 1 *Mithqal* = 7 carats.
12 *Mithqals* = 420 carats therefore 1 *Mithqal* = 35 carats.
165 *Mithqals* = 1150 carats therefore 1 *Mithqal* = approximately 6.8 carats.

Robert Webster valued 1 *Mithqal* at 39.40 carats.[10]

The *Encyclopedia of Islam* mentioned 14 1/2 *Mithqals* = 320 carats, therefore 1 *Mithqal* = approximately 22 carats.[11]

In my opinion, the value determined by Drs. Hassan and Khafaga in their study of al Tifaschi's book (which corresponds to that of Ahmad al Qalqashandi, a legal scholar and secretary in the Mamluk chancery in the fourteenth century,[12] and also to that of Clément Mullet, one of the handful of Western orientalists to have studied and translated al Tifaschi's work[13]) is the correct one. This value was estimated as follows[14]: 1 *Mithqal* was equal to 24 carats; 1/6 *Mithqal* was equal to 4 carats, 1 *Mithqal* was equal to 5.088 g, therefore 1 carat was equal to 0.222 g and one grain to 0.053 g.

As to the dinar, Clément Mullet estimated it to be worth French Francs 14.586[15] while M. Bauer estimated 2 Dinars at Sterling Pounds 6.[16]

Prices of pearl necklaces in the year A.D. 1240[17]

Al Tifaschi explained that in Baghdad, the regular pearl necklace was made of thirty-six pearls, the lightest necklace weighing 1/6 *Mithqal* (or 4 carats).

		Dinars
10 necklaces of 1/6 *Mithqal* each (4 carats each), total 40 carats	=	3/4
10 necklaces of 1/4 *Mithqal* each (6 carats each), total 60 carats	=	1
10 necklaces of 1/3 *Mithqal* each (8 carats each), total 80 carats	=	1 1/4
10 necklaces of l/2 *Mithqal* each (12 carats each), total 120 carats	=	2
10 necklaces of 3/4 *Mithqal* each(l8 carats each), total 180 carats	=	4
10 necklaces of 1 *Mithqal* each (24 carats each), total 240 carats	=	10
10 necklaces of 1 1/4 *Mithqals* each (30 carats each), total 300 carats	=	15
10 necklaces of 1 1/2 *Mithqals* each (36 carats each), total 360 carats	=	20
10 necklaces of 1 3/4 *Mithqals* each (42 carats each), total 420 carats	=	25
10 necklaces of 2 *Mithqals* each (48 carats each), total 480 carats	=	35
10 necklaces of 2 1/4 *Mithqals* each (54 carats each), total 540 carats	=	40
10 necklaces of 2 1/2 *Mithqals* each (60 carats each), total 600 carats	=	50
10 necklaces of 2 3/4 *Mithqals* each (66 carats each), total 660 carats	=	70
10 necklaces of 3 *Mithqals* each (72 carats each), total 720 carats	=	80
10 necklaces of 3 1/4 *Mithqals* each (78 carats each), total 780 carats	=	90
10 necklaces of 3 1/2 *Mithqals* each (84 carats each), total 840 carats	=	110
10 necklaces of 3 3/4 *Mithqals* each (90 carats each), total 900 carats	=	150
10 necklaces of 4 *Mithqals* each (96 carats each), total 960 carats	=	200

Al Tifaschi pointed out that if pearls are of exceptional quality and orient, the price of one single necklace weighing 4 *Mithqals,* for example, can reach 30 dinars, that is, 300 dinars for 10 necklaces.

He quoted the following prices for individual necklaces made of large pearls:

		Dinars
1 necklace of 4 1/2 *Mithqals* (108 carats)	=	40
1 necklace of 4 3/4 *Mithqals* (114 carats	=	55
1 necklace of 5 *Mithqals* (120 carats)	=	65
1 necklace of 5 1/4 *Mithqals* (126 carats)	=	80
1 necklace of 5 1/2 *Mithqals* (132 carats)	=	85
1 necklace of 5 3/4 *Mithqals* (138 carats	=	90
1 necklace of 6 *Mithqals* (144 carats)	=	100
1 necklace of 7 *Mithqals* (168 carats)	=	150

The above prices could go up sharply according to the quality of the pearls and could also go up (according to the above-mentioned proportion) with the increase in weight.

Prices of individual pearls

Jean Taburiaux stated that the price of pearls depends on the following criteria: shape, quality, color, sheen, and orient.[18] These are exactly the same five criteria proposed by al Tifaschi, namely: shape (in terms of sphericity), referred to by the word *dahradja*; size, referred to by the word *kibr al djirm*; quality of color, referred to by the word *naqa'* (meaning purity and freedom from defects affecting the color); sheen, referred to by the word *basiss*; and orient, referred to by the word *ishraq.*[19]

In addition to the above criteria, al Tifaschi mentioned that the type of hole drilled in the pearl affects its shape and consequently its price. As pearls were drilled with unsophisticated, simple instruments, the size and shape of the hole were apparently important criteria in the valuation of pearls, as an excessive loss of weight or an irregular piercing could negatively affect the value and beauty of a pearl.

Prices quoted by al Tifaschi for best quality pearls are the following: (Sterling Pounds estimates according to M. Bauer's calculations of 1895 are also given for reference).[20]

• One pearl weighing 1 *Mithqal* (24 carats or 96 grains): 300 dinars (STG 900)

• If two twin pearls of 1 *Mithqal* each are available, the price for both can reach more than 700 dinars (STG 2,100)
• Two twin pearls weighing together 1 *Mithqal* (24 carats or 96 grains): 100 dinars (STG 300)
• Two twin pearls weighing together 2/3 of a *Mithqal* (16 carats or 64 grains): 27 dinars (STG 81)
• Two twin pearls weighing together 1/2 a *Mithqal* (12 carats or 48 grains): 20 dinars (STG 60)
• Two twin pearls weighing together 1/3 of a *Mithqal* (8 carats or 32 grains): 5 dinars (STG 15)

Supposed medicinal benefits and superstitions

• If powdered and added to a collyrium, it strengthens the optic nerves.
• It regulates heart beat and prevents heart attacks; this is the reason why doctors use pearls as an ingredient in their medicines for the heart.
• It chases away fear and anguish.
• It stops hemorrhages.
• It polishes teeth to a beautiful white color.
• If powdered and mixed with fat, it acts as an antidote against poison. Nowadays, Hindus, Chinese, and Japanese people eat powdered pearls; being composed of pure calcium carbonate, they supply the body with a precious vitamin.

Notes

1. *Encyclopedia of Islam*, 250.
2. Al Tifaschi, *Azhar al Afkar fi Djawahir al Ahdjar*, 3.
3. Ibid., 7.
4. Al Biruni, *Al Djamahir fi Ma'arifat al Djawahir*, 127.
5. Ibn al Akfani, *Nukhab al Zakha'ir fi Ahwal al Djawahir*, 29.
6. Bauer: *Precious stones*, vol. 2, 58.
7. Al Tifaschi, 7.
8. Ibid., 6.
9. Sinkankas, *Emerald and other Beryls*, 546-547.
10. Webster, *Gems: Their Sources, Descriptions, and Identification*, 939.
11. *Encyclopedia of Islam*, vol. 3, 528
12. Hassan and Khafaga, commentary, 210.
13. Mullet, "Essai sur la Minéralogie Arabe,"505.
14. Hassan and Khafaga, commentary, 210-211.

15. Mullet, 505.
16. Bauer, *Precious Stones*, vol. 2, 257.
17. Al Tifāschi, 7-9.
18. Taburiaux, *Pearls: Their Origin, Treatment, and Identification,* 207.
19. Al Tifaschi, 9.
20. Bauer, vol. 1, 257.

CHAPTER 5

RUBY, SAPPHIRE, AND THE OTHER CORUNDUM VARIETIES (*YAQUT, YAQUT ASMANDJUNI*, AND *YAWAQEET UKHRA*)

AL TIFASCHI used the word *Yaqut* to describe ruby, sapphire, and the rest of the corundum family. Nowadays also, most Arabs refer to ruby as *Yaqut,* sapphire as *Safeer* and the rest of the corundums either as "white sapphire, yellow sapphire," and the like or "white ruby, yellow ruby," and the like. Al Tifaschi mentioned ruby immediately after pearl, as it was considered the most valuable gem after pearl at that time. He devoted a long chapter to this stone and mentioned several points of great importance which I shall describe at length due to their extraordinary implications in modern gemology.

Ruby and the other corundums are aluminum oxides. The true color of pure corundum is white, but different impurities present in the stone give it different colors, thus chromium gives ruby its beautiful red color, iron and titanium give us blue sapphire.

During al Tifaschi's time, ruby came from a mine called Sahran (or Sahbaran or Sahayran?) on an island (the name of which was not given) "behind" Sri Lanka, at a distance of 40 *Farsakh* (that is 240 km) from Sri Lanka, and of an approximate superficy of 62 *Farsakh* (372 km). This could have been Borneo or perhaps Burma, as al Biruni described this island as "not really being an island, as it is very close to the shore and almost attached to it."[1] However, later on in the text, al Tifaschi mentions the formidable Djabal al Rahun as the "mountain of rubies," from which wind and rain carry the rubies away. Thus, they are found as placer deposits or as waterworn pebbles, in the beds of rivers surrounding this mountain. This Djabal al Rahun he says, is the one where Adam landed after having fallen from paradise. As a matter of fact, there is a mountain in Sri Lanka, called Adam's Peak, around which rubies abound in river beds. Ratnapura (or city of rubies) lies near the foot of this mountain, which, according to local legend, is indeed where Adam landed after being expelled from paradise. Al Biruni had also mentioned Djabal al Rahun as the mountain where Adam first set foot on earth, describing the peculiar vegetation around the place where Adam's footprints left their mark as growing in a way that resembles the shape of a person kneeling in reverence.[2] This Djabal al Rahun is none other than Adam's Peak, and visitors who go up this mountain can see the imprint of a footmark, covered in glass for protection, which Buddhists say is the footprint of Buddha and is a well-known pilgrimage site in Sri Lanka.

At any rate, what is interesting here is that al Tifaschi mentions the legend of the "mountain of rubies." He was told by people who had visited Sri Lanka that the people of this island revert to the following trick to obtain these stones if winds and rain fail to bring them as usual during certain years. He wrote: " For the mountain where corundums are found is a formidable one, difficult to reach and inaccessible on the top. Many eagles live and build nests on top of this mountain, as it is uninhabited by mankind. Therefore, the people of this country slaughter an animal, skin it, and cut it into large pieces, then leave it at the foot of the mountain. Eagles come to retrieve this meat and carry it to their nests, and if they place it on the ground, the corundum stones stick to it. Then other eagles gather around the meat trying to steal it, and the stones fall off due to their heavy weight, to be picked up by the people." [3]

It is worth noting that R. Webster also mentions an old Burmese legend where rubies are found in an inaccessible valley, and therefore the locals threw lumps of meat in it so that the vultures would go and retrieve them, picking up at the same time the precious stones that had adhered to the meat.[4] Pliny had mentioned the legend of the "valley of the diamonds,"[5] which was also mentioned by al Tifaschi (see in the "diamond" chapter). Therefore, it seems there were two different legends related to diamond and ruby, both originating from the Far East and both well known in ancient and medieval times. Al Tifaschi also mentioned that some ruby crystals were found "recently" in the village of Tura, in the Gizeh district in Egypt, and that the chief mineralogist of the sultan at his time, who was in charge of the emerald mines in Egypt, had shown him some of these red stones.[6]

Al Tifaschi noted that, as corundum is heavy in weight, heavier than any other stone that resembles it, it is found in alluvial deposits in river beds after being transported by wind and rain. Both statements are true, as the specific gravity of ruby is 3.97- 4.01, making this stone the heaviest after zircon (with a specific gravity of 4.0 to 4.69). Also, al Tifaschi's use of the wording "alluvial deposits" is indeed the first reference in gemology to the occurrence of corundum in secondary or placer deposits.

Colors and varieties

In the West, ruby and sapphire were discovered to belong to the same species less than 200 years ago.[7] However, al Tifaschi knew for a common fact already 750 years ago that ruby and sapphire are members of the same family of stones, together with other colored varieties. He grouped them together under the name "ruby" (*Yaqut*) and divided them as follows[8]:

Yaqut ahmar (ruby), with the following variety of colors:
• *Wardi* color, of shades varying from rose red to pinkish white. The darkest color of this group is similar to that of a red rose (*ward*), the lightest having a pale pink shade almost reaching white.
• *Khamri* (wine-colored), of deep red mixed with purple. The best color of this group is similar to that of the cheirantus (gillyflower), the lightest being of a purplish-white shade.
• Red color: The best is of a deep red similar to that of the reddest safflower; the lightest being of a pale safflower color.
• *Summaqi* (porphyry) color, of dominating black shade
• *Bahramani* color: Of an intense red, the lightest shade of which is that of the reddest safflower. This variety was the best and the most expensive one. Al Kindi had described it as having a color similar to that of blood from a major vein on a clean, highly polished, pure silver tray.

Yaqut asmandjuni (sapphire), with the following variety of colors:
• Light-blue color
• Sky-blue color, described as being similar to the color of lapis lazuli and considered to be the best one. Even today, this particular deep, even, blue shade is called "cornflower blue" and is considered to be the best color for sapphire.
• Ultramarine color
• Navy-blue color
• *Sannuri,* of a greyish-blue shade, considered to be the worst kind of blue sapphires.
• *Zayti* color, or greenish-yellow (the color of oil), was also considered to be a poor quality of sapphire.

Yaqut akhdar: Green corundum

Yaqut asfar: Yellow corundum, with the following variety of colors:
- *Khaluqi*: Yellowish-orange color, similar to that of turmeric
- Transparent light-yellow color, with strong brilliance
- *Gulnari*: Orange-pink color, similar to that of the flower of the pomegranate tree. This shade was the most expensive of the yellow to orange corundums.
- *Utrudji*: Citrine-yellow color
- *Tibni*: Straw-yellow color
- *Mishmishi*: The color of apricots

Yaqut abyad (white corundum): Considered then–and today–to be the least valuable of all the corundums.

White sapphire was divided into two kinds: the *maha'i*, which displays the strongest luster, best transparency, and is colorless, free of impurities; therefore it is the best kind; and the other lesser kind, which was called the "male." This is a reference to the curious "male and female" theory of stones, which probably started before Theophrastus but was first reported by him in around 300 B.C. However, it differs in this context: Theophrastus reports that the "male" variety of any stone was always of stronger or darker color than the "female" one. As for Pliny, "male" stones were those with more brillance and luster, while "female" ones had a duller brilliancy and more subdued luster.[9] For al Tifaschi, the "male" variety was less valuable and had less luster, transparency and other qualities than the *maha'i* one. The word *maha'i* here does not refer to the "female" kind, but to a kind that resembles the *maha*, or the completely colorless, pure white glass.

Al Tifaschi described ruby as being the hardest of all stones after diamond, giving it a hardness of 3 on his own scale, right beneath diamond, to which he gave a hardness of 4. (Ruby and diamond occupy, respectively, numbers 9 and 10 on the Mohs scale of hardness). He also specifically mentioned and stressed that, although ruby is a very hard stone, it can easily (and only) be cut by diamond, as "wood is by iron," therefore correctly assuming that, although ruby is only second on the scale of hardness after diamond, its hardness is far inferior to that of diamond. Al Tifaschi described the luster and brilliance of ruby as being exceptional, which no other colored stone can match. In fact, the great hardness of corundum gives it a luster that can almost resemble that of diamond. Another interesting observation made by al Tifaschi refers to cleavage in ruby. Al Tifaschi observed that cleavage in ruby is not really cleavage; he called it " false cleavage," or *shibh tashqiq* (the exact translation of his words would be "pseudocleavage"). Gemologists agree today that the type of cleavage present in ruby is called "pseudocleavage" or "parting," which is breaking along directions parallel to twinning planes due to lamellar twinning in ruby.

According to Ibn Masaweih, whom al Tifaschi quoted on this point, a ruby stone of 5 *Mithqals* (120 carats) was extremely rare, while sapphire can occur in sizes of 40 *Mithqals* (960 carats). Al Kindi had mentioned that he had seen in the coffers of *Khwarizm* (Persia), a sapphire of 60 *Mithqals* (1440 carats) carved into the shape of a slave girl sitting, clutching her knees against her chest between her interlocking fingers, her chin resting on her knees.[10]

Inclusions

Inclusions in stones are numerous and depend generally on locality and conditions affecting the growth of the crystal.

Al Tifaschi mentioned one of the first references in gemology to types of inclusions in gemstones and their subsequent treatment. He accurately described some of these inclusions to be air, water (or other liquid), solids, or even some kind of worms who die upon contact with the air. He also described another type of inclusion common in ruby, resembling "holes similar to worms in wood." By this he probably meant rutile needles that are frequent inclusions in rubies and sapphires from Sri Lanka. These impurities, says al Tifaschi, affect the color, brilliance and value of the stone, hence they are removed by special processes or treatments.

The first kind of treatment is used to remove air, water and some kinds of solids in more or less liquid states (he used the word *teen* which means "mud," to describe these liquified solids): A drill is used on the stone to allow these impurities to go out.[11] By that, al Tifaschi probably meant either that the drill is used like a vacuum pump to remove the impurities, which, being liquid or air, can easily slide out of the stone through the tiniest of channels or evaporate; or that the drill makes a tiny hole in the stone to let the impurities out. The Arabic word he used was *mithqab,* which literally translates as "hole maker." As mentioned earlier in the chapter on diamonds, drills were often made of iron or lead, and diamond "points" were used to cut or drill holes in stones. It was also customary to drill hooves on the sides of stones for attachment in necklaces or pendants.

The second kind of treatment is used to remove the "wormwood-like" cavities. Here the drill was placed on the side of the inclusion. No further details were given on this process. The above two processes were mentioned only by al Tifaschi; no old or modern gemological textbook mentions such processes.

The third and most important kind of treatment, according to al Tifaschi, is used to remove "the blackish tinge from ruby," as well as to produce color in white corundum or improve that of other corundums. This is the notable heat treatment process.

Al Tifaschi was not the first one to mention heat treatment of corundum. He himself quotes Aristotle on several occasions on this subject. But he was the first one to describe in detail the heating process for the purpose of gem improvement.

Heat treatment of sapphire and ruby has been increasingly practiced in the West since the 1940s. Sapphires from Sri Lanka and Australia are heat-treated to improve their shades, but little is known about heat treatment of rubies. In his book *Identifying Man-made Gems*, M. O'Donoghue described several processes of heat treatment for the development of asterism in a stone, for the removal of unwanted rutile from a stone, for the diminution of banding, etc., in addition to the following process, which I shall quote here for its resemblance with the one al Tifaschi described in the thirteenth century. This process involves lightening the color of too dark stones or changing stones of undesirable colors into better ones: "Heating the stone for an extended period in an oxidizing atmosphere may lighten the color . . . .A colorless stone may be the result of a long period of heating. Where a yellow color is possible, heating may produce green or yellow; and where purple sapphire with a chromium content is heated, ruby may be the final product."[12] Technical information including degree of heat and chemical reactions, were given, but not actual information on "how" the heating process takes place. Apparently this had been a long-kept secret by the locals.

However, al Tifaschi described the process of heat treatment accurately and in detail. This treatment was mentioned only in the ruby chapter, therefore we can assume that it was carried out only on corundum. Al Tifaschi described the following process, which was used to remove the blackish tinge from rubies:

> Stones from the mother rock where ruby is found are taken, powdered, dissolved in water, then pounded until they are sintered together. They are used to cover the stone (to be treated) completely, then they are placed on a rock (crucible), other rocks are gathered around it, and lots of wood is added on top. (The fire is lit) and is kept alive by constant blowing (with bellows) and adding of more logs of wood (according to the amount of black in the stone), which they know from careful observation and experience.[13]

A1 Tifaschi explained that the people who specialize in heat treatment of ruby in Sri Lanka are well experienced in the matter, and they know exactly how much wood to burn; how strong the fire should be and how long the treatment should last, which he says could be anywhere between one hour to twenty consecutive days and nights. These people take turns in blowing on the fire to keep it constantly alive day and night, until the undesirable black hue is removed, and a beautiful color is obtained. Other corundums were also treated in the same way; al Tifaschi explains that only rubies are stable when heated, that is, they do not change color. This is confirmed scientifically, as oxidation does not affect stones colored by chromium (like ruby). The other corundums all change color upon heating, often becoming white. However, the careful heat treatment of such stones can make them resemble the more valuable varieties, and al Tifaschi mentions that he saw many stones being sold in Cairo as blue and yellow sapphires, while he knew that they were in fact white sapphires that had been treated to become blue and yellow.

How could this scholar, whose main occupation was that of a judge and a poet, have obtained such thorough knowledge of gemstones as to be able to differentiate between heat-treated and natural sapphires at a time when no gemological instruments whatsoever were available to facilitate this task? Even now, the most experienced gem tester or jeweler is unable to identify a heat-treated stone without the use of accurate instruments to support his examination, and in some stones the treatment is so well done and stable that a decision is indeed impossible to make.

Al Tifaschi described another test used to differentiate between ruby and other stones similar to it, namely, resistance of the stone to intense heat. He explained that ruby does not decrepitate or calcinate when subjected to intense heat, unlike, for instance, emerald, which crumbles and calcinates due to its brittleness and low hardness.[14]

Due to its ability to withstand intense heat without being affected, al Tifaschi said that ruby is a "hot" stone, giving it number 4 on his heat scale, the other corundums occupying a decreasing order on this scale according to their color (e.g., blue sapphire 3, orange, pink, and yellow 2, with white sapphire the coldest of the corundums at number 1). He also pointed out that ruby can withstand a quick drop in temperature upon cooling without being affected, unlike all others stones. Heat treatment can serve as an additional test to differentiate between ruby and other red stones, he added, as, if any red stone changes color upon heating, it is definitely not a ruby, but some other gemstone similar to it or some manufactured (imitation) one.

From the above, it is clear that heat treatment has been practiced for a very long time. Ibn Masaweih had already mentioned that the color of ruby is improved by the heating process, and al Kindi used to heat-treat rubies himself in order to improve their color. Therefore, one can assume that heat treatment was routinely practiced by Arab "gemologists" for at least 1200 years, while in India and Sri Lanka, this process was known to exist since Aristotle's times.

Polishing

The method used for polishing ruby, explains al Tifaschi, differs from that of other stones, as ruby can only be polished on a copper disk (or lap), using powdered and ignited agate mixed with some water. Ignited agate probably meant a microscopic break-down of the fine structures in agate (chalcedony) and a resulting fine-grained powder when the calcined material is crushed (and probably flotation-refined).

Prices

Al Tifaschi gave the following average prices of ruby, "according to two marketplaces, namely Baghdad and Cairo, as they represent the standard to which all others refer:"[15]

- A stone weighing 1/2 *Dirhem* (or 8 carats, as 1 *Mithqal* is equal to 24 carats and equal to 1 *Dirhem* and 3/7[16]) = 6 Egyptian gold *Mithqals*; each carat was worth 10 silver *Dirhems.*
- A stone weighing 1 *Dirhem* (16 carats) = 32 Dinars; each carat at 2 dinars.

• A stone of 1 *Mithqal* (24 carats) = 60 dinars; each carat at 2 1/2 dinars.
• A stone weighing 1 1/3 *Mithqals* (32 carats) = 96 dinars; each carat at 3 dinars.
• A stone weighing 2 *Mithqals* (48 carats) = 192 dinars; each carat at 3 1/2 to 4 dinars.

But, al Tifaschi hastened to add, these prices can go up sharply with the quality of the stone. For example "the price of a good stone weighing 1 *Mithqal* could reach 100 gold *Mithqals*, if the stone is a *bahraman* ruby of top quality, color, brightness, and brilliance; well ground, polished, and worked, thus having lost a lot of its original size."[17]

Blue and greenish-yellow sapphires were valued at 4 dinars/*Dirhem*; yellow sapphire at 2 dinars/*Dirhem;* and white sapphire at 1 dinar/*Dirhem.*

An interesting remark that al Tifaschi made when describing prices of corundum is related to modern economic theories: According to him, the value of all stones depends on two things: one inherent in the stone itself and the other independent of the stone. The inherent factors include the quality of the stone and its size; the external or independent factors depend on supply and demand for this stone in the market and rarity of the stone, as well as the location of its mines. For example, if the stone is found in a remote region of a faraway land, its value will automatically increase in those countries the furthest from the country of origin. Therefore, in addition to the valuable gemological information he gave us, al Tifaschi also advanced marketing and economics theories such as availability of the commodity, rarity of the material, supply and demand affecting its value and price, upon which many of our modern financial and economics theories are based.

Supposed medicinal benefits and superstitions

• Ruby strengthens the heart (meaning the self-assurance) of the wearer and his standing amongst people. It also increases respect of the people for him.
• As a talisman, it prevents heart palpitations and recurrent anxieties.
• Lightning never strikes the wearer of a ruby.
• Ruby (or any other corundum variety) was never seen in the hand of a drowned man.
• Ruby prevents hemorrhage as well as blood clotting if worn on a necklace.

Notes

1. Al Biruni, *Al Djamahir fi Ma'arifat al Djawahir*, 44.
2. Ibid., 44.
3. Al Tifaschi, *Azhar al Afkar fi Djawahir al Ahdjar*, 15.
4. Webster, *Gems: Their Sources, Descriptions, and Identification*, 78.
5. Tolansky, *The History and Use of Diamond*, 18.
6. Al Tifaschi, 29.
7. Schumann, *Gemstones of the World*, 86.
8. Al Tifaschi, 17.
9. Caley and Richards, *Theophrastus on Stones*, commentary, 124.
10. Al Biruni, 76.
11. Al Tifaschi, 16.
12. O'Donoghue, *Identifying Man-made Gems*, 68.
13. Al Tifaschi, 16-17.
14. Ibid., 19.
15. Ibid., 21-22.
16. Hassan and Khafaga, commentary, 210.
17. Al Tifaschi, 22.

CHAPTER 6

EMERALD *(ZUMURRUD)*

EMERALD is a beryllium aluminum silicate, the most valuable stone of the beryl group, which also includes aquamarine (blue), morganite (pink), heliodor (yellow), goshenite (colorless), and precious beryl (yellow shades, pink, and colorless). There is also a strawberry-red variety of beryl.

The name emerald is derived from the Greek *Smaragdos,* which came originally from the Semitic *Zabardjad* (also *Zabargad, Zabardadj* and *Zabardag*). Up to the year A.D. 1000, this word referred to all green-colored stones, including peridot and emerald. However, by the twelfth century, *Zabardjad* referred exclusively to peridot, and emerald was known as *Zumurrud.* It is often mentioned that "*Sabardschah, Sbaragd, Semerid, Siberget, Tabarget, Zamarut, Zamorrute, Zamorat*" are all Arabic names of emerald.[1] Probably some confusion arose from the Arabic name of peridot, namely *Zabardjad* (or the other above-mentioned pronounciations); as peridot was confused with emerald, and the name for peridot was used also to describe emerald, together with the word *Zumurrud* (or *Zumurrudh*). In fact, emerald is exclusively known as *Zumurrud* or *Zumurrudh* in Arabic, and any other name is in fact a mispronounciation or an error.

Al Tifaschi described the following types (or colors) of emerald[2]:

1. *Zhubabi*: The best kind, with an even, deep-green color and good transparency. It is named *zhubabi* because of its color, which resembles that of the cantharides fly (in Arabic *zhubab* means flies).
2. *Rihani*: of a basil-green color
3. *Silqi*: of a chard-green color
4. *Sabuni*: of a soap-green color

The best color was the *zhubabi,* which was the most valuable, and to which the medicinal benefits were exclusively attributed. The best variety of emerald was one with a whitish tinge, called *'Arabi* (the Arabian). It was found in Hidjaz, Saudi Arabia.

The location of the famous Egyptian mines, the Cleopatra mines, was lost until 1818[3] when Caillaud "rediscovered" them. In fact, al Tifaschi had referred to these mines in his book as located in a mountain at the borders of Egypt and Sudan, behind Aswan, extending like a bridge. Drs. M. Y. Hassan and M. B. Khafaga, respectively professor of geology, dean of the School of Sciences; and head of the Geology Department at Al Azhar University, Cairo, Egypt, in 1977, confirmed this location to be schist formations in the eastern range of mountains sandwiched between the Red Sea and the Nile River.

The chief mineralogist of the sultan explained to al Tifaschi that the principal emerald mine was in Wadi al Shah, between the two towns of Qus and Aizab, at a walking distance of ten days from Qus. The mining sites were called Makamin, Takayu, Qu' Ghazi, and Wadi al Nahr, the latter being the biggest mine. Wadi al Shah was so named because a copper statue of the shah was found there. The chief mineralogist also explained to him how emeralds are found: "We find friable talc where emeralds are found embedded in a soft red earth . . . .We could hit the vein, thus taking the emerald crystal whole . . . as for the small stones, they are found by sieving the sand."[4]

Al Tifaschi used the words *qasab* and *qasaba* to describe the rough emerald crystal still embedded in its matrix. The word *qasab* or *qasaba* in Arabic translates here as "elongated crystal" or "reed" and describes perfectly the form of a rough emerald crystal. Al Tifaschi also stated that emerald is found in talc, which is a black stone that yields golden marcasite when heated. According to him, emeralds are never free of incipient cracks. This is indeed true, as emerald is brittle, and the stones are usually full of inclusions (which could be other crystals, empty cavities, cavities filled with air and gas, or solid and gas, or other). When dendritic forms are seen, the French term *jardin* is used. As emerald has indistinct cleavage parallel to the basal pinacoid crystal shape, it exhibits tiny cracks along these directions, which also mar the appearance of the stone. Al Tifaschi warned that a stone called *mash* (an unidentified stone that could be green beryl or green tourmaline) resembles emerald in color, weight, and brittleness, and is found together with emerald; therefore, it can be confused with it. In fact, if we compare the characteristics of tourmaline and emerald, we find that green tourmaline (called verdelite) can resemble emerald in color; its specific gravity (3.02-3.26) is heavier than that of emerald (2.67-2.78); while its hardness (7-7 1/2) is similar to that of emerald (7 1/4-7 3/4). It also crystallizes in the trigonal system, which some gemologists consider as a division of the hexagonal system to which emerald belongs. Cut green tourmaline could, therefore, as al Tifaschi pointed out, easily be confused with emerald. Precious beryl of green color can also be confused with emerald, especially that it belongs to the same beryl family. However, al Tifaschi was quick to point out that one can easily differentiate between the emerald and its imitations with the foil test: The color of emerald improves noticeably when it is foiled, while that of *mash* darkens almost to opaque black or turns yellowish.

Foiling, which consists of using a thin sheet of gold, metal, or other colored foil at the back of the stone in order to improve, lighten or darken its color, was apparently routinely practiced at that time. In fact, al Tifaschi advised that this procedure could be used to differentiate between genuine stones and their imitations, thus providing an additional test for gemstone identification. Foiling is still practiced nowadays, but it is mostly for the purpose of deceiving the purchaser into believing that the stone is of a better quality than it really is. It consists of exactly the same procedure described above, the stone usually being set in a closed setting to avoid detection.

Al Tifaschi also stated that, apart from the inclusions of which emerald is virtually never free, another defect usually found in emerald is the difference in color shades found in different parts of the same stone. He was probably referring either to strongly dichroic emeralds that show different shades of color when viewed from different angles, or to color zoning in emeralds, which are due to changes in chemical composition during growth. Thus, emeralds sometimes show a light green or almost colorless core, surrounded by darker colored zones.

Another defect that al Tifaschi noted in some emeralds is their deformed shape. As a matter of fact, schist-type deposits, such as the Egyptian ones, often yield deformed crystals.[5] He warned that stones that could be confused with emerald include green jasper, green jade (nephrite), peridot, and green sapphire.

A special and unique power attributed only to emerald of the best quality was its ability to make snakes' eyes bulge and almost come out of their sockets. The unfailing test to differentiate between best quality emeralds and its imitations was the following: "I have read about this characteristic of emerald in books of philosophers, then tested it myself and found it to be true. For I had a pure *zhubabi* emerald that I wanted to test using the eyes of snakes. So I hired a man to catch a snake, which he did, and I put it in a large basin. I took a piece of wax, stuck it at the end of an arrow and glued the emerald on it, then I approached it from the eyes of the snake, which was leaping at the arrow, moving furiously trying to get out of the basin. As the arrow neared its eyes, I heard a light noise (as if one was killing a nit on his nail), and I saw that the eyes of the snake had bulged considerably and that it roamed perplexed in the basin, not knowing which direction to turn to nor wanting to get out. Most of its movements became still, and it had stopped leaping altogether."[6]

It would indeed be interesting to try this test nowadays to ascertain the quality of emeralds.

Prices

Prices of best quality emeralds (*zhubabi* kind) were as follows[7]: A stone of one *Dirhem* in weight (16 carats) was worth 4 dinars/carat = 64 dinars. Prices double according to size but are not significantly lower for smaller emeralds–unlike other stones–due to the noble and exceptional benefits of emerald.

As for the other varieties (or colors), their value is negligible due to their lack of benefits. If we compare prices with ruby, we note that a best quality emerald was worth double the price of a good ruby (for a stone of 16 carats). However, the price of an exceptional, well-cut, top quality ruby was more than that of the best emerald.

Supposed medicinal benefits and superstitions

The following applies only to best quality emeralds, as it was common belief in those days (perhaps based on Indian lore, or vice versa as the home of the emerald in ancient times was exclusively Egypt) that the better the quality of the stone, the greater the benefits[8]:

- If one stares at it for a long time, it can improve eyesight.
- Kings and rulers order that emeralds be worn by their children, as this stone guards against epilepsy.
- If worn as a talisman over the liver and the stomach, it can cure dysentery.
- It can cure stomach pains.
- Poisonous animals do not approach the wearer of an emerald.
- It can be worn on a necklace for good luck and to guard against evil.
- If worn over the thigh of a woman in labor, it can speed up childbirth.

Notes

1. Sinkankas, *Emerald and other Beryls*, 623-627.

2. Al Tifaschi, *Azhar al Afkar fi Djawahir al Ahdjar*, 25.
3. Webster, *Gems: Their Sources, Descriptions, and Identification*, 100. Also, Bauer, *Precious Stones*, vol. 2, 311.
4. Al Tifaschi, 29.
5. Sinkankas, 275.
6. Al Tifaschi, 26.
7. Ibid., 28.
8. Sinkankas, 64.

CHAPTER 7

PERIDOT *(ZABARDJAD)*

PERIDOT is a magnesium iron silicate, belonging to the mineral species olivine, together with the lighter-green colored variety.

Very little was known about peridot in the thirteenth century. As a matter of fact, al Tifaschi was the first to point out that peridot was a separate species from emerald; at least until the times of al Biruni, they were considered to be the same stone. Al Tifaschi mentioned that this stone was not found like other minerals in a host rock or a mine, but that, until the time of writing his book in the year A.H. 640 (A.D. 1240) the few stones available were the ones found in Alexandria, Egypt, by digging archaeological sites and old ruins. It was believed that they were the remnants of Alexander the Great's treasures. Al Tifaschi did not mention that peridot was obtained from St. John's island (also called *Zabardjad* or *Zabargad* island) in the Red Sea, which is the only location to yield large crystals suitable for cutting. This location must have been unknown at the time, until the beginning of nineteenth century, when the Viceroy Mohammad Ali of Egypt had these mines worked exclusively for him.

Al Tifaschi described three shades of peridot: a dark green, a light green and a medium green of good brilliance and luster, the latter being the best and the most valuable. He added that this stone was highly esteemed in India and in Europe.

Prices

A stone of 1/2 *Dirhem* or 8 carats was worth 1 dinar.[1] From this low price, we note that peridot was not considered to be a very valuable stone.

Supposed medicinal benefits and superstitions

- It improves eyesight (probably by staring at it).
- If worn as a talisman, it bestows purity and peace of mind on the wearer.

Notes

1. Al Tifaschi, *Azhar al Afkar fi Djawakir al Ahdjar*, 30.

CHAPTER 8

SPINEL (*BALAKHSH*)

SPINEL is a magnesium aluminum oxide and comes in a variety of colors, including red, pink, violet, yellow, orange, mauve, blue, and black. According to al Tifaschi, the different colors of spinel are red, pink, orange, yellow, and green.

This stone used to come from Balakhshan (Afghanistan or Northern India?), hence its name. Al Tifaschi mentions that this stone is a hard one, although less so than ruby. He did not mention cracks or cleavage, which indeed are not present in spinel. According to R. Webster[1]: "In olden times red spinel was often known under the name 'Balas ruby,' a name devised most probably from Balascia (or Badakhshan) in northern India, [from] where the earliest stones were said to have come." In my opinion, "Balas" could have been derived from the Arabic word *Balakhsh*, which was itself derived from Balakhshan, as the early Arab gemologists knew that this stone was different from ruby, while in the West, and until the nineteenth century, "Balas ruby" referred to a variety of ruby that was in fact a spinel. According to Ball,[2] the earliest reference to the Badakhshan mines of spinel was in A.D. 951. However, the fact that al Biruni mentioned that Nasr ibn Ya'qub[3] (pre-tenth century Persian gemologist) reported on this gem-producing site proves that it was well-known even before that period.[3]

Two famous spinels are in the English crown jewels; one is an oval stone in the Imperial State Crown and is known as the "Black Prince's Ruby," and the other is a 361-carat stone known as the "Timur Ruby" on which are engraved the names and dates of some of its previous owners. These stones were thought to be rubies until the nineteenth century. Other famous spinels are the spinels of the Wittelsbacher's crown of 1830, which were also thought to be rubies until recently.[4]

Spinel was recognized to be a stone separate from ruby only 150 years ago[5]; before that it was classified under ruby, as it also occurs with it. However, al Tifaschi knew that spinel was a different species and described this stone with its different colors, giving the various characteristics that separate it from ruby and saying that the only benefit of this stone (medicinal or otherwise) is that it closely resembles ruby in its color, appearance, luster and transparency. Indeed, 600 years before Western gemologists acknowledged that spinel was different from ruby, al Tifaschi had already established this fact. What is even more stunning is that in his book *Al Djamahir fi Ma'arifat al Djawahir,* al Biruni had already mentioned that spinel was thought to be ruby until the times of Al Bouayh (the Buyids), that is until the tenth century. It was then discovered to be a different stone, and its price was dropped accordingly.[6]

Al Tifaschi mentioned three varieties of spinel: red, green, and yellow, the best being the red variety.[7] Some jewelers differentiated between five varieties of spinel: The *mu'aqrabi* of a bright red color, the *'atashi* of a duller red shade, the *inari* of a pomegrenate color, the *niaziki* of a lesser red than the *inari*, and finally the yellow variety, similar in color to the yellow corundums. As yellow spinel is very rare, al Tifaschi could have been referring to a variety of zircon, including it among the spinel varieties.

Prices

Al Tifaschi said that a good quality spinel was worth half the price of a ruby of similar quality.[8] If we take the example (under the ruby chapter) of a stone weighing 1 *Dirhem* (or 16 carats), its value was 32 dinars or 2 dinars/carat. That of a spinel of similar quality and weight would therefore be 16 dinars or 1 dinar/carat.

Notes

1. Webster, *Gems: Their Sources, Descriptions, and Identification*, 32.
2. Ball, *Historical Notes on Gem Mining*, 720-721.
3. Al Biruni, *Al Djamahir fi Ma'arifat al Djawahir*, 82.
4. Schuman, *Gemstones of the World*, 100.
5. Ibid., 100.
6. Al Biruni, 82.
7. Al Tifaschi, 31.
8. Ibid., 32.

CHAPTER 9

GARNET *(BANAFSH,* also Called *MAZENDJ* or Almandine Garnet*)*

GARNET is the name of a group of stones, which includes the following species: almandine, pyrope, spessartine, andradite, uvarovite, and grossular. Hessonite and tsavolite are gem varieties of grossular garnet; demantoid and topazolite are gem varieties of andradite. As garnet is a very complex group of gemstones (covering more than twenty known varieties), with an even more complex chemistry, I shall simply add here that five commercial types are available in gem qualities: pyrope, almandine, spessartine, grossular, and andradite.

Very little was written by al Tifaschi on garnet, probably as this stone was not considered to be a very valuable one. He only mentioned that garnets and spinels are formed the same way and that they are found in the same deposits. In fact, both stones belong to the cubic (isometric) system and are indeed found in the gem-bearing gravels of Sri Lanka.

Al Tifaschi described the following varieties of garnets:

1. *Mazenbi*: A transparent stone of light red color, which was considered to be the most valuable of the garnets. The stone in question was probably pyrope. As a matter of fact, Ibn al Akfani did mention 100 years after al Tifaschi that *mazenbi* stones came from Sri Lanka, India, and *Bilad al Afrandjah,* meaning Europe.[1]
2. A dark red stone
3. A violet garnet, described as being "of black color tinted with a bluish-red shade."[2] By this stone was probably meant a pyrope-almandine type of garnet, of violet-purple color.
4. *Azbazashtt*: Light orange-yellow variety, by which was probably meant hessonite. It is worth mentioning that hyacinth zircon is similar in color to hessonite (reddish-yellow) and, before the use of refractive index and double refraction measurements, the two stones were often confused with one another. Zircon was known even before al Tifaschi's times, as al Biruni called it *Kharjun* and described it as being of a paler red than

rubellite (tourmaline).[3] Therefore, it is highly unlikely that the two stones were confused by al Tifaschi, especially in view of the much heavier specific gravity of zircon when compared with that of garnet.

5. Al Tifaschi also described another stone, which he called *Mazendj*, in the tourmaline chapter. This stone was of a deep red color with a violet-blackish tinge, so dark that its color and transparency only showed when it was hollowed out from the underside. This stone was probably almandine garnet, which is of an intense dark red color, and which even nowadays is often cut in the *cabochon* style with its back hollowed out. (This procedure makes the stone thinner, thus allowing more light to penetrate, which lightens its color).

It is interesting to note that al Tifaschi described the varieties of garnet under one chapter and heading, namely *Banafsh*, while he classified *Mazendj* as a separate stone, similar in color to rubellite (red tourmaline), of such a dark shade that it needed hollowing out for its true color to appear. That he described almandine separately attests to its relative importance and separation from the rest of the species. As a matter of fact, almandine, the name of which is derived from Alabanda, a town in Turkey, was considered to be an important stone throughout the ancient and medieval periods. The "Alabanda" stone was called "carbuncle," and from the times of Theophrastus and Pliny until the early eighth century, the town of Alabanda was supposed to be the source of these "fiery" gems. However, neither al Kindi, (ninth century), al Biruni (tenth-eleventh century), al Tifaschi (thirteenth century), nor Ibn al Akfani (fourteenth century) mention Alabanda as a source of garnet. In fact, the only sources for these stones were reported by all to be India and Sri Lanka. Only Ibn al Akfani noted that pyropes also came from Europe. Therefore, the Alabanda source must have been only a cutting and trade center and not a mining area, as suggested by J. D. Rouse.[4] However, one cannot help but notice the similarity between the Arabic name, *al Mazendj,* and the Western one, almandine.

Prices

Garnets were not considered to be very valuable stones. Al Tifaschi gives an idea of their price as follows: The best variety (*mazenbi*) was worth a quarter of the price of a spinel of similar quality or even less. A *mazenbi* stone of 1 *Mithqal* (24 carats) was worth 2 dinars. The red variety was worth half the price of the *mazenbi* (or 1 dinar per *Mithqal*); the *azbazashtt* variety was worth half the price of the red variety (or 1/2 dinar per *Mithqal*), and the violet type was worth half the price of the *azbazashtt*, or 1/4 dinar per *Mithqal*.[5]

If we take the example (see the spinel chapter) of a spinel of 16 carats valued at 1 dinar/carat, the price of a garnet of equal quality, according to al Tifaschi, would be 1/4 dinar/carat. This would make it 6 dinars/*Mithqal* and not 2 dinars/*Mithqal* as he mentioned above.

However, the important point here is that garnet was considered to be much less valuable than spinel, as is the case today, This price was far lower than it was in the West at that time, where garnets were equal to rubies in value as late as in the seventeenth century (in De Boodt's time).[6] Almandine, however, was considered to be a separate stone from ruby and garnet and was cheaper.[7]

Supposed medicinal benefits and superstitions

Only *azbazashtt* (hessonite) was supposed to stop nose bleeding when hung externally (presumably over the nose).

Notes

1. Ibn Al Akfani, *Nukhab al Zhakha'ir fi Ahwal al Djawahir,* 18.
2. Al Tifaschi, *Azhar al Afkar fi Djawahir al Ahdjar,* 32.
3. Al Biruni, *Al Djamahir fi Ma'arifat al Djawahir,* 90.
4. Rouse, *Garnet,* 63.
5. Al Tifaschi, 32.
6. Rouse, 42.
7. Ibid., 53.

CHAPTER 10

TOURMALINE *(BADJADI,* also called *BIJADI, BAZADI, BIZADI)*

TOURMALINE is a complex boro-silicate that is found in Sri Lanka, East Africa, Southwest Africa, Brazil, Madagascar, and the United States of America. Tourmaline surpasses all other stones with its variety: It comes in all colors, as well as in multicolored varieties. Some tourmalines are blue on one end and pink on the other; others have a red core surrounded by green (watermelon stone).

The few scholars who have studied and researched Arab writers on the subject of gemstones have different opinions as to the nature of the stone called *Badjadi.* Drs. M. Y. Hassan and M. B. Khafaga are of the opinion that by *Badjadi,* al Tifaschi meant almandine garnet. Both Clément Mullet and Julius Ruska are of the same opinion. R.V. Dietrich[1] reports that the Arabic name for tourmaline is *Turmala.* However, having searched al Biruni, al Tifaschi, and Ibn al Akfani, I found no such word in use. Neither did the earlier jewelers and "specialists" on gemstones mentioned by al Biruni describe any stone called *Turmala,* although al Biruni did describe a stone (in the *Balakhsh*–or spinel–chapter) which could have been tourmaline or particolored corundum, as follows: "The Razi brothers say that the same stone could display all the colors and that they had actually seen one that was multicolored and was red, yellow, green, blue, and white at the same time, and they knew that heating could remove all the colors except the red one, which was stable."[2]

As for al Tifaschi, he described this stone as follows: "*Badjadi* is a stone of wine color, without brilliance except in few specimens, these resembling ruby. It is colder and softer than ruby, and if the stone is found in its host rock, it is dark and opaque; only when cut by the specialists do its color and beauty appear, and it becomes brilliant. The best is the one of strong red color, with good brilliance. It does not shine unless it is foiled, and its underside is hollowed out[3]. . . A stone that resembles *Badjadi* is [called] *Mazendj,* which is a deep red stone with a tinge of black, softer than *Badjadi.*" It could be differentiated from *Badjadi* by the clearer, stronger luster of the latter, and the fact that "it does not pick up any dust particles from the floor." Later on in the text, al Tifaschi continues the description of this stone as follows[4]: "One of the characteristics of *Badjadi* is that when it is rubbed against the hair of the head or of the beard, then placed on the floor, it picks up straw and other objects." The above description leads us to believe that by *Badjadi* was meant tourmaline, while *Mazendj* referred to almandine garnet (see the garnet chapter).

Al Tifaschi mentioned that almandine was less hard than tourmaline; in fact, the hardness of both stones on the Mohs scale is very similar, with that of almandine garnet at 7 1/2 and that of tourmaline ranging from 7 to 7 1/2.

Pyroelectricity is a peculiar effect exhibited by some stones. When heated, they develop a positive charge at one end of the crystal and negative at the other, thus attracting small pieces of paper or ashes or dust from the floor. Many stones other than tourmaline exhibit this phenomenon, namely amber, quartz, and almandine garnet. It is reported that "W. Gilbert in the sixteenth century found this property was shared by other substances" (that is, other than amber), "and he named the effect 'electrification' after electron, the Greek word for amber."[5] The discovery of the pyroelectric property of tourmalines was made by John Canton in 1754 and Frank Ulrich Aepinus in 1757.[6] However, al Tifaschi had described this property as early as 1240, and pyroelectricity had been mentioned even earlier in the works of other Arab scholars, as al Biruni compares the pyroelectric effect of amber to that of *Badjadi* (tourmaline), which means that this effect was observed in tourmaline long before it was discovered in amber, at least in theArab world.[7] As a matter of fact, one of the earliest references to the pyroelectric property of tourmaline was made in *Kitab al Ahdjar li Aristatalis* (Aristotle's Book on Stones). It is believed that this book was not written by the famous Greek philosopher, but by another author using this name. It is unknown whether this author is Greek or other; however, if it was indeed written by Aristotle, then this effect was already well known by 330 B.C. This book was translated into Arabic by Luqa Ibn Serafion (or Asrafion?) before the tenth century, as al Biruni often quoted passages from this pseudo-Aristotelian work in his book *Al Djamahir fi Ma'arifat al Djawahir*, and J. Ruska estimates the date of the book to be somewhere before the middle of the ninth century. Therefore the pyroelectric effect of tourmaline was well known by at least the ninth century in the Arab world.

Tourmaline was found together with ruby, in the island "behind"

Sarandeeb (Sri Lanka). Tourmaline was and is still abundant in Sri Lanka and its pyroelectric properties were well known by the locals. The Dutch, who first imported it in the early eighteenth century, called it "*aschentrekker*" or ash puller, and used it to pull ashes out of their pipes.[8] *Badjadi* was probably rubellite, the red variety of tourmaline.

Prices

A batch of stones weighing 1 *Mithqal* (24 carats) was worth 1/2 dinar.[9]

Supposed medicinal benefits and superstitions

• If worn in a ring, tourmaline can dispel nightmares.

Notes

1. Dietrich, *The Tourmaline Group*, 1.
2. Al Biruni, *Al Djamahir fi Ma'arifat al Djawahir*, 74.
3. Al Tifaschi, *Azhar al Afkar fi Djawahir al Ahdjar*, 33.
4. Ibid., 33.
5. Webster, *Gems: Their Sources, Descriptions, and Identification*, 876.
6. Sinkankas, *Emerald and other Beryls*, 49.
7. Al Biruni, 210.
8. Dietrich, 168.
9. Al Tifaschi, 34.

CHAPTER 11

DIAMOND *(ALMAS* or *MAS)*

DIAMOND is called *Almas* in Arabic. It is generally accepted that the name is derived from the Greek *Adamas,* which means unconquerable (note the similarity between the two words). But could "*al Mas*" for Arabic have preceded "*Adamas*" for the Greek, as it is more likely that the Arab world would have known about Indian diamonds–probably the earliest known–before the Greeks?

Diamond has a unique chemical composition among gemstones, being composed of a single element, pure crystallized carbon. India was the only known source of diamonds in the old times until the Brazilian fields were discovered in the eighteenth century and the South African ones in the nineteenth century. In Europe, Marco Polo wrote about the riches of India at the end of the thirteenth century. It is believed that the Portuguese Garcias do Horto was the first to record the diamond mines of India in 1565, one century before Tavernier's noted travels.[1]

Al Tifaschi quoted Ibn Masaweih (ninth century) when he wrote about the legend of the valley of the diamonds in India. According to Ibn Masaweih, diamonds came from a deep valley in India. This valley was so deep that no one could reach it; therefore people used to cut tender meat into pieces and throw it in this valley, where eagles would go and retrieve them. Diamonds would adhere to this meat, and when the eagles flew carrying it, other eagles attacked them, and the diamonds would fall off the meat to be picked up by the waiting locals. This legend was already recorded by Pliny (A.D. 25-79) in volume 37 of his *Natural History*, which was devoted to precious stones, and which could have been the source of Ibn Masaweih's account, unless, of course, he had heard it firsthand and reported it independently. At any rate, there was an obvious truth in this legend, as diamonds do adhere to pieces of raw meat, having an affinity for grease. As a matter of fact, one of the modern techniques for sorting out diamonds from other stones consists of running these stones on a conveyor belt covered with thick vaseline. Diamonds stick to this belt, while other stones fall off under the light pressure of the water, which diamonds repel. According to R. Webster,[2] the West discovered this property of diamonds only recently, when F. Kirsten noticed that diamonds are water-repellent.

Diamond was not considered to be as valuable as pearls and rubies in the thirteenth century and before. As a matter of fact, al Biruni mentions that, during his time, diamond was found in many colors, including red, green, grey, black, yellow, and white and that people in Baghdad (Iraq) and Khurassan (Persia) did not differentiate between the values of the different colors, as diamond was used there only for drilling stones, due to its great hardness, and for poisoning!!![3] Apparently, diamond was powdered into as fine grains as possible and hidden in salt or sugar to kill the person intended by tearing at his insides.

Al Tifaschi mentioned two colors of diamond: "the *zayti* one and the *billawri* one; the *zayti* being the most valuable. The *billawri* variety is of a strong white color, similar to rock crystal; the *zayti* is a whitish color mixed with a tinge of yellow similar to the color of oil, and resembling pharaonic glass."[4] It seems that color played the most important role in evaluating gemstones, as even diamonds with a yellowish tinge were more valuable than colorless ones. The most highly prized stones, according to al Tifaschi, are the ones that display such strong fire that they can actually project this light, "similar to a rainbow," on their surroundings. All transparent stones can exhibit dispersion, this phenomenon being visible especially in colorless stones. The different colors of white light (the colors of the rainbow) are refracted at different angles, and the rays that are reflected within the diamond return to the eyes of the beholder split into the spectral colors. Al Tifaschi described this phenomenon accurately when he mentioned: "There is a variety of diamonds with such magnificent brilliance that they can project this brilliance on a nearby wall, piece of cloth, or face of a person or other objects, by producing a light of various colored rays similar to a rainbow."[5] Diamonds that did not display this "fire" were distributed to the trade for cutting corundum.

From al Tifaschi's remark that lower quality diamonds were used for cutting corundum, we note that diamond points were used for drilling stones and diamond powder was used for grinding and polishing stones, especially corundum, diamond being the only material harder than corundum. S. Tolansky[6] had mentioned that the Arab scholar Ibn Manzur recorded the extreme hardness of diamond by saying that if a diamond is squeezed in an iron vise, it will penetrate the iron and itself remain unaffected. The fact is that Ibn Manzur had obtained most of his information on gemstones from al Tifaschi, as explained by the Arab scholar Abdel Qadir Zamamah, who wrote on the life and achievements of al Tifaschi.[7]

Al Tifaschi goes on describing the exact crystal habits of diamond: "A characteristic of diamond is that all of its varieties have fixed angles: six angles" (meaning an octahedron shape, which has six corners or points), "eight angles" (meaning a cube, which has eight angles) "or more" (meaning dodecahedron and other).[8] And in another place: "Its angles are surrounded by surfaces of triangular shape . . . . If it breaks, it breaks only in triangular shapes, even if broken into the smallest of parts."[9] The above is probably the first reference in gemology to the crystalline forms of diamond, well before the seventeenth-century works by Steno, Keppler, Boyle, and others.[10] Diamond belongs to the cubic or isometric system and often occurs in octahedron form (described as being "two equilateral four-sided pyramids base to base."[11] Another form in which diamond occurs is the cube, but this is rare. Other forms include the dodecahedron (twelve faces) and the icositetrahedron (twenty-four faces). However, as India was the exclusive source of diamonds in the old times (apart perhaps from Borneo), diamonds from this region occur mostly in octahedrons and dodecahedrons, which correspond to the shapes described by al Tifaschi.

Cleavage

The art of cleavage must have been known for a very long time but was probably kept secret in view of the remarkable applications of such a knowledge. When al Tifaschi mentioned that diamond breaks only in triangular shapes, even the smallest of its parts,[12] he was probably referring to cleavage in diamond, which effectively occurs along directions parallel to the octahedral faces (and therefore are triangular in shape). He certainly did not mean "break" in the sense of "being fractured" or "powdered," as he refers to powdering diamonds when he describes the usual test that was used for diamonds at that time, which consisted of hitting it with an iron hammer on an iron anvil. He also did not mean fracture, as he goes on describing a method for cleaving diamond: According to him, diamond is the hardest of all stones; it can break or scratch all stones as well as erase all engravings from their surface while it is itself extremely hard to break. The only way to break it (meaning to cleave it) was "to put it inside a piece of resin, then place it inside a wicker reed; then it is lightly and carefully tapped with a hammer made of lead, making sure that the iron does not touch the diamond, or it is placed in a tube of lead, and is hit carefully by some hard instrument, then it can be broken."[13]

The first European record of cleavage is by De Boodt in 1604, followed by Tavernier in the record of his second trip to India in 1638.[14] Although he described the above method for cleaving diamond, al Tifaschi dwelled at length on the extreme hardness of diamond, saying diamond is "so hard that it can scratch or cut all other stones . . . and any stone that comes into contact with it will be deeply scratched . . . . It cannot be broken by any stone as it softens and weakens them all. Another characteristic is that it can drill holes into pearls, corundum, emerald, and other stones, as well as any stone that cannot be worked with iron; this (is done) by placing on one end of a piece of wood a piece of iron on top of which a stone of the size wanted to make the appropriate hole is placed."[15] Al Tifaschi gave diamond a hardness of 4 and coolness of 4 on his scales (the highest numbers).

Prices

Al Tifaschi estimated the price of a 1 carat diamond at 2 dinars (about 6 Sterling Pounds according to M. Bauer's estimate of 1896).[16] He also quoted al Kindi as saying that he (al Kindi) had inspected stones varying in size between the tiniest piece (the size of a mustard seed) to the largest (the size of a walnut). Al Kindi had also stated that he had never seen a stone larger than a walnut in size (an approximately 100-carat octagonal stone). He estimated that the average prices of diamonds in Baghdad were as follows (prices in Sterling based on Bauer's estimates):

- The highest price for a batch of diamonds weighing in total 1 *Mithqal* (24 carats) was 80 dinars (STG 240).
- The cheapest price for a batch of diamonds weighing in total 1 *Mithqal* was 15 dinars (STG 45).
- An individual stone of 1/2 *Mithqal* (12 carats) suitable for a finger-ring was rare, and its price was double that of the mustard seed size, triple that of the *filfila* (or *fulful*) size, quadruple that of the following size, and so on.[17] This system was probably based on the now-obsolete Indian rule, whereby the price of a diamond exceeding 1 carat is reached by squaring its weight in carats and multiplying by the price of the 1 carat stone. For example, if the price of a 1-carat stone is STG 2, that of a 4-carat stone would be 4 x 4 x 2 = STG 32.[18]

However, al Tifaschi was not the first one to give a price list for diamond (as believed by some Western gemologists). Al Biruni had established several price lists for gemstones in the beginning of the eleventh century in his book *Al Djamahir fi Ma'arifat al Djawahir*. Some were even in tabulated form (pearls and emeralds), giving prices per units of weight and how many pieces per units of weight; others were mentioned in the individual chapter of the relative gemstone.[19] Al Biruni quoted the price of 100 dinars for a stone of 1/2 *Mithqal* (12 carats). He also quoted al Kindi's earlier estimate (as did al Tifaschi later on) which was as follows: "A batch of stones of the *falafel* size weighing in total 1 *Mithqal* of best quality diamonds that showed a strong dispersion is worth 80 dinars. I did not see a stone bigger than a walnut, and the price of such a stone appreciates three to five times above that of the smaller sizes."[20] Al Biruni had also quoted the two famous brothers al Razi as saying they did not see a stone larger than 3 *Dirhems* in weight (48 carats). They also stated that "the price of a batch of stones weighing in total 1 *Dirhem* (16 carats) is worth 100 dinars; and an individual stone of a similar weight is worth 1,000 dinars."[21]

Supposed medicinal benefits and superstitions

Al Tifaschi quoted some supposed medicinal benefits attributed to diamonds by Aristotle and other old scholars; therefore, I shall refrain from mentioning them here, as they were not his own.

He pointed out that if a person swallows even the tiniest piece of diamond, it tears his intestines and kills him instantly. But he also reported that other persons are not in the least affected when swallowing diamonds; and he concluded that both possibilities were correct as some diamonds were still impregnated by the venom from the snakes of the valley where the diamonds were found, while others were not.

Notes

1. Bauer, *Precious stones*, vol. 1, 153.
2. Webster, *Gems: Their Sources, Descriptions, and Identification*, 34.
3. Al Biruni, *Al Djamahir fi Ma'arifat al Djawahir*, 95.
4. Al Tifaschi, *Azhar al Afkar fi Djawahir al Ahdjar*, 36.
5. Ibid., 36.
6. Tolansky, *The History and Use of Diamond*, 19.
7. Hassan and Khafaga, commentary, 16.
8. Al Tifaschi, 36.
9. Ibid., 36.
10. Bauer, 120.
11. Webster, 16.
12. Al Tifaschi, 36.

13. Ibid., 107.
14. Tolansky, 34-35.
15. Al Tifaschi, 38.
16. Bauer, 257.
17. Al Tifaschi, 39.
18. Bauer, 105.
19. Al Biruni, 97.
20. Ibid., 97.
21. Ibid., 97.

CHAPTER 12

MOONSTONE CAT'S EYE (*'AIN AL HIRR*)

MANY people in the Middle East today think that by cat's eye are meant the quartz varieties of tiger's eye (brownish-yellow) or hawk's eye (bluish-brown), which display this peculiar effect. However, there are many stones that display chatoyancy, including apatite, beryl, chrysoberyl, diopside, enstatite, quartz, scapolite, and tourmaline and therefore can be called "cat's eye" with the prefix of the relative stone, that is, tourmaline cat's eye, quartz cat's eye, etc.

The description that al Tifaschi gives of this stone is the following: "This stone is a strange one as the overwhelming color is white with an astonishing radiance [coupled] with a delicate transparent clarity, but a bluish spot can be seen inside it, similar to a cat's eye catching the light in its moving pupil; this spot moving constantly . . . . If it is moved to the right, the spot moves to the left, and vice versa, likewise upwards and downwards; it is exactly like the eye of a cat, and for this reason it was called cat's eye. If the stone is broken or is cut into tiny pieces, this spot will appear in every single one of these pieces.[1] "If the stone is held still . . . water similar to moving waves is seen inside it."[2]

Al Tifaschi was very meticulous in his descriptions of all stones, and by the one mentioned above, he must have meant either quartz cat's eye or moonstone cat's eye, the best variety of which is white with a bluish sheen. Both gemstones are abundant in Sri Lanka, but the color and schiller of moonstone, being bluish (and not white as in quartz cat's eye) fits al Tifaschi's description better. Al Biruni had mentioned a stone called *Hadjar al Qamar,* which translates exactly into "moonstone," and by which was probably meant the moonstone we are familiar with today. As al Tifaschi was familiar with al Biruni's book and often quoted him in *Azhar al Afkar fi Djawahir al Ahdjar*, he must have known about *Hadjar al Qamar,* but he did not relate it to his description of *'Ain al Hirr* due to the absence of chatoyancy in the plain moonstone. Also, the fact that al Tifaschi did mention that this stone was not described in any old or new book on gemstones confirms the nature of this stone as moonstone cat's eye, as no mention of such a stone was made in the writings of either al Kindi or al Biruni, whose works preceded al Tifaschi's.

Moonstone is a silicate of potassium, sodium, and aluminum, consisting of layers of orthoclase and albite feldspar, which give it the special adularescence for which it is known. By moonstone cat's eye is meant moonstone that displays chatoyancy, that is, when cut *en cabochon*, it shows a wavy mobile band of light gliding across the dome at the slightest movement. Al Tifaschi informed us that people in India and Sri Lanka held this stone in great esteem and that people in India revered this stone as they revered statues.

Prices

Al Tifaschi said that the price of this stone depends on demand for it. In the Arab countries, a stone weighing 1 *Mithqal* was worth only 5 dinars, but in India and Sri Lanka, this stone was more valuable.[3] Some moonstone cat's eye stones fetched very high prices. He gave the example of a stone of 1 *Mithqal* that a man called Isma'il al Salameh bought for 700 gold *Ya'qubi* dinars in India in A.H. 641 (A.D. 1241). When this man died, his heirs sold the stone to Ibn Rasul, the ruler of Yemen, for the price of 7,000 *Maliki* dinars, and still it was said that this price was too low for such a fine and rare specimen.[4]

Supposed medicinal benefits and superstitions

- It protects from the evil eye and all types of evil, human or spiritual.
- The stone offers the same benefits as those of ruby (see the ruby chapter) in addition to the following: One is that it safeguards its owner's money and protects him from all mishaps and dangers; the other is that, if worn by a person in times of battle and defeat, and if this person pretended to be dead and lies amongst the cadavers in the field, his enemies see him as a corpse drowned in his own blood and therefore leave him be; and thus his life is spared!

Notes

1. Al Tifaschi, *Azhar al Afkar fi Djawahir al Ahdjar*, 39.
2. Ibid., 40.
3. Ibid., 41.
4. Ibid., 41.

CHAPTER 13

BEZOAR *(BAZAHR)*

AL TIFASCHI went into lengthy descriptions of some stones, while mentioning others only by passing. Thus, we see long chapters devoted to pearl, ruby and sapphire, diamond, emerald, moonstone cat's eye, lapis lazuli, and coral; while the remaining stones were mentioned rather briefly. We can therefore assume that the above-mentioned stones were the most important and valuable ones at that time.

However, the lengthiest description belongs to one stone that is nowadays unfamiliar to us, namely, bezoar. The most precious of gems, according to al Tifaschi, was pearl, and pearl is a secretion of the oyster. Coral was also considered to be a valuable gemstone, and coral is the axial skeleton of the coral polyp, a primitive type of plantlike animal. Therefore, it is not surprising to see another organic material so highly considered and esteemed, especially as this stone was supposed to have an almost miraculous power: It acted as a powerful antidote against all types of poison.

This is what al Tifaschi explained about this strange stone: The name *Bazahr* comes from the Persian *Bak* (meaning cleanser) and *Zahr* (meaning poison), hence the name of the stone, "cleanser of poisons." According to Clément Mullet, the French word *bézoard* has its roots in the Arabic name.[1] There are two types of bezoar, one mineral and the other animal.

The mineral variety comes from China, where large stones are found and made into knife hilts and other ornaments. It is a soft, brittle stone, with a white streak powder. Colors are white or yellow with tiny specks of other colors. The best variety is yellow with green specks.

Supposed medicinal benefits of the mineral variety

- It is used as an antidote specifically (and only) against scorpion bites.
- If powdered and mixed with water and spread over an ecchymosis, no more than twenty-four hours are needed for the wound to be completely healed and the swelling gone.

The animal variety (which according to al Tifaschi is the most interesting and the most valuable), is a soft, brittle stone, of spotted yellow color and white streak powder. It is composed of thin successive layers. Animal bezoars are in fact calculi found exclusively in deer that are abundant in the border regions between China and India. The favorite food of this animal is poisonous snakes, especially their young.

There are three theories on the formation of this stone:

1. According to some, this stone is formed inside the eyes of the deer. After eating all these poisonous snakes, the deer feels a rash over his entire body, which he quenches by immersing himself in water up to his head. Steam then slowly rises from his body until it reaches his eyes where it solidifies upon contact with the air and remains stuck on the hairs between his eyes and nostrils. And so on, other stones are formed in the same way and get stuck one over the other until they become too heavy and fall or too bulky so the deer gets rid of them by rubbing his head against a stone or a tree.
2. Another theory is that the stones were formed inside the hearts of deer but al Tifaschi does not believe this was the case.
3. The third theory, which al Tifaschi favours, is that the stone is formed within the gallbladder of certain deer, and indeed the bitter taste of bile can clearly be tasted when licking the stone.

Prices

A small stone was worth 1 dinar.[2] Al Tifaschi pointed out an important aspect of the jewelry trade at that time, as he told the story of an *A'djami* (i.e, Persian) merchant who brought eighteen bezoar stones for sale in Alexandria. He gave them to the *Dallal* (broker) who in turn took them to the *Amin al suq* (market supervisor) who confirmed that they were genuine bezoar stones. When the stones reached the shop of al Tifaschi's friend (whom he described as a merchant from al Andalus, Spain, who had established business in Alexandria); he picked up only two stones out of the whole batch, as he recognized that the others were not genuine.
We can note several interesting points from the above story:

1. Spanish jewelers or merchants often established business in the Arab countries, and especially Alexandria, which was a well-known trading center at that time.
2. Foreign merchants who wanted to sell stones could not go about their

business directly themselves. They had to go through intermediaries, called *Dallal* and *Amin al suq* (broker and market supervisor, respectively) to dispose of the stones. Both the *Dallal* and the *Amin al suq* had to give their opinion as to the genuineness of the stone.
3. There were competent jewelers and "gemologists" who could, by sheer firsthand observation, recognize genuine from fake stones.

Supposed medicinal benefits of the animal variety

- If the stone is rubbed against the spot bitten by a poisonous animal (or plant), it remains stuck until all the poison has been removed from the body and then falls off by itself.
- Drinking a little powdered bezoar (either mixed with olive oil or water) can extract the poison from the body through perspiration.

Al Tifaschi wondered about the mysteries of the Creation, whereby two opposites meet in the same animal: On the one hand, the powerful antidote against poison (bezoar) is found in the gallbladder of deer; on the other, a powerful poison is extracted from the knotted end of the tail of this same animal. This poison, al Tifaschi explains, acts slowly, painfully, and irreversibly, killing by freezing the blood in the body, in a sleeplike death. Many kings and rulers of the time commissioned hunters to provide them with this poison. Al Tifaschi went on, saying that it is indeed a wonderful act of the Creator that both poison and its antidote run in the blood of the same animal without harming him, while at the same time its meat is completely safe for human consumption and constitutes a succulent and substantial meal. He had apparently tasted this meat himself on several occasions and enjoyed it greatly. Poison was found in the tails of all deer, in the Orient (North Africa and the Middle East), as well as in the Far East. However, the best bezoar stones came specifically from deer of the Chinese-Tibetan border.

Gemologically speaking, bezoar cannot be considered as a gemstone, as it lacked two of the three basic requirements of a gemstone: beauty and durability. However, its rarity and outstanding medicinal value had earned it a respectable place amongst those gems considered to be the most important and valuable during al Tifaschi's time. One should not forget that this scholar lived at a time when superstitions and legends were part of daily life, and the value of a gemstone increased steadily with the value of its (alleged) benefits.

In the bezoar chapter, al Tifaschi also described the "rain-bringing stones." These stones were apparently rubbed against one another, resulting in clouds gathering and heavy rain falling. Special people practiced this art, and the rulers took them on their trips during the summer to provide them with a convenient occasional shower whenever needed.

Notes

1. Mullet, "Essai sur la Minéralogie Arabe," 144.
2. Al Tifaschi, *Azhar al Afkar fi Djawahir al Ahdjar*, 55.

CHAPTER 14

TURQUOISE *(FAIRUZADJ)*

TURQUOISE is a hydrous copper aluminum phosphate found in hot, arid regions of the world. Nowadays, the main sources of jewelry turquoise are in the Southwestern United States, mostly from copper deposits in Arizona, while production from Iran and Egypt is nearly nil. China has also developed an important high-quality turquoise deposit. Al Tifaschi mentioned only Nishapur as the source of gem quality turquoise. This is strange, as he must have certainly known about the Egyptian sources, which were famous as early as 3000 B.C., especially as he lived most of his life in Egypt, let alone his extensive and sound gemological and geographical knowledge.

The logical explanation is therefore to assume:

• That only Persian turquoise was considered of gem quality. Even today, turquoise stones from Iran are the most highly prized, and their prices exceed by far those from Egypt or the United States. A particular kind commonly known as *Abdul Razzaq* is supposed to be the best, being of very dense, high-grade quality, not porous, thus, it does not change color due to absorption of foreign substances. Persian turquoise is often free of veins (traces of black or brown limonite in the stone showing as black veins inside the blue matrix), while Egyptian turquoise often shows these veins, which are considered as a sign of genuineness in the stone.

• Egyptian turquoise was a very common stone, used and worn by common people at that time, therefore al Tifaschi did not mention it.

The second explanation seems to be the correct one, as both al Biruni (tenth-eleventh century) and Ibn al Akfani (fourteenth century) also do not mention Egyptian sources of turquoise in their books on gemstones. However, a contemporary of al Tifaschi, Mohammad ibn Manzur, (A.D. 1232-1311) wrote a treatise on mineralogy, mentioning the rich turquoise mines of Egypt. According to him, there was a legend that told that the richest mines were called the Isaac mines, as they were supposedly opened by Isaac, the son of Abraham.[1] Ibn Manzur described seven varieties of turquoise, according to their degree of hardness and their shade of color. The best variety was named after Abu Ishaq (the father of Ishaq, or Abraham); the worst was the *'andelibi,* of a weak light-blue color.[2] Ibn Manzur was in fact a great admirer of al Tifaschi and well acquainted with his writings. He copied a lot of al Tifaschi's ideas and work and even went to great lengths to obtain the only copy of the type of encyclopedia that al Tifaschi wrote under the title of *Fasl al Khitab fi Madarek al Hawass al Khams li Uli 'al Albab* (Final Information Given by the Five Senses to People Who Have Minds [or Intelligent People?]) and recopied the best parts in poetry form under the title *Nithar al Azhar fil' Layl wal' Nahar* (meaning roughly On Petals Shed by Flowers Concerning Day and Night).

Al Tifaschi stated that copper enters in the composition of this stone, and that there were two varieties of turquoise: one of a "pure radiant blue color" called *bushaqi* and another less valuable variety called *qadjandji.*[3] He did not elaborate on this inferior variety, but we can assume that he could have meant the poorer quality stones that changed color to green, or the ones showing traces of limonite, or even chrysocolla, which can sometimes be confused with turquoise.

Blue turquoise only is considered valuable in North Africa, the Middle East, and Europe and America, while green turquoise is the one that is prized and more valuable in China and Tibet.

An important point to mention here is the description al Tifaschi gave of the genesis of this mineral: "Turquoise is a coppery stone formed by copper vapors rising from its host rock."[4] Although turquoise is not formed in this way, being a basic phosphate of aluminum with a little copper and iron, it is interesting to note here what is one of the first mentions in gemology of the hydrothermal theory of the formation of minerals.

The fact that turquoise owes its color to copper was confirmed by R. Webster in the twentieth century, who stated that "the cause of the beautiful blue colour of turquoise is not known for certainty. The sky-blue perfection colour, known in America as 'robin's egg-blue,' has been ascribed to copper."[5]

The color of turquoise is affected by grease, perspiration, and musk, says al Tifaschi, and this stone is highly prized by the Berber princes of Morocco for use in finger rings and adornment of weapons.

Prices

Prices varied a lot according to quality. The average price ranged from 1 dinar for a good stone suitable for a ring to 1 *Dirhem* for a lesser quality one.[6] Berber princes paid as much as 10 Moroccan dinars for a good turquoise.

Supposed medicinal benefits and superstitions

- It improves eyesight if one stares at it for a long time.
- If powdered and added to a collyrium, it is beneficial for the eyes.

Notes

1. Bauer, *Precious Stones*, vol. 1, 394.
2. King, *The Natural History of Gems or Semi-precious Stones*, 63.
3. Al Tifaschi, *Azhar al Afkar fi Djawahir al Ahdjar*, 56.
4. Ibid., 56.
5. Webster, *Gems: Their Sources, Descriptions, and Identification*, 242.
6. Al Tifaschi, 57.

CHAPTER 15

CORNELIAN (CARNELIAN) *('AQIQ)*

CORNELIAN is a translucent variety of chalcedony. The word *'Aqiq* comes from *'Aqqa,* meaning to cleave, to split open; and it is believed that cornelian was called *'Aqiq* because of its toughness and ability to cleave other stones. 1 Al Tifaschi grouped many chalcedony varieties under the name of *'Aqiq,* saying that this stone came in five colors: Deep red, flesh red (the colors of cornelian); blue, white (the colors of chalcedony), and black (the color of onyx). The best variety was the deep red, followed by the flesh red, which shows light whitish banding. The remaining varieties were not considered of any importance.

Al Tifaschi also said that this stone was a hard one (he did not specify the scale of hardness, but this stone has a hardness of 7 on the Mohs scale, therefore his assumption that it was a hard stone is correct). He also described this stone as being "hot" (in relation to his heat and cold scales). Cornelian used to come mostly from Yemen, where it was distributed to other countries.

Prices

Cornelian was mostly used in finger-rings, one ring selling for 4 *Dirhems.* A good stone that was engraved sold for 1 *Dirhem,* while large ones fashioned into knife and dagger hilts sold for 1 dinar or slightly less.[2]

Supposed medicinal benefits and superstitions

- The deep red variety used as a talisman controls fear in times of battle.
- The flesh red variety was supposed to stop hemorrhage from any part of the body.
- If any variety of this stone is rubbed against the teeth, it polished them and removed tooth decay. It could also prevent gum bleeding.

Notes

1. Ibn al Akfani, *Nukhab al Zakha'ir fi Ahwal al Djawahir*, 78.
2. Al Tifaschi, *Azhar al Afkar fi Djawahir al Ahdjar*, 58.

CHAPTER 16

AGATE (*DJAZ'*)

AGATE is microcrystalline (cryptocrystalline) quartz and occurs in various colors, in layered and banded forms. W. Schumann gives the following varieties of agate: eye agate, layer agate, dendritic agate, enhydritic agate, fortification agate, moss agate, scenic agate, pseudo-agate, tubular agate, sard stone, brecciated agate, and others.[1]

Al Tifaschi said that agate was found in many varieties, naming six different types, including the *habashi* (Ethiopian), *baqarawi* (from *baqara*, cow) *ghurawi* (?), *farissi* (Persian), *'assali* (honey-colored) and *mu'arraq* (banded) types.[2] The two most important varieties according to him are the *baqarawi*, which is banded red and white (nowadays the dark red/white color of this variety is known as sardonyx and the flesh-red/white one as cornelian-onyx); and the *habashi*, which is banded black and white (known today as banded onyx). Al Tifaschi described sardonyx as being composed of three layers: A red opaque one, followed by a white opaque one, then by a translucent colorless one. As for onyx, he described it as being composed of a white layer sandwiched between two black ones. Al Tifaschi explained that agate was polished by hand with honey on and improved noticeably when cooked in oil. Agate used to come mainly from Yemen and China in the thirteenth century. Al Tifaschi said that this stone is one of the most resistant and difficult stones to work with, giving it a hardness of 2 and a degree of coolness of 2 on his scales. As a matter of fact, agate has a hardness of 6 1/2-7 on the Mohs scale and is very tough; "it is used in industry because of its toughness and resistance to chemicals."[3]

Prices

A good quality bead weighing 1 *Mithqal* (24 carats) sold for 2 *Dirhems.*[4]

Supposed medicinal benefits and superstitions

Al Tifaschi explained that this stone was called *Djaz'* (which means fear in Arabic) because it could arouse fear in the heart of people. He quoted old scholars as saying that in Yemen, only ignorant people wore jewelry decorated with agate, as they ignored the fact that this stone brought problems and provoked nightmares. It was also known at that time that Chinese people had a deep aversion for this stone for the same reasons.

The rare benefits of this stone are the following:

- If placed near a woman who just gave birth, it soothes and diminishes her pains.
- It can cauterize open purulent wounds and stop hemorrhage.
- It is used for polishing ruby, as previously explained.

Notes

1. Schumann, *Gemstones of the World*, 114.
2. Al Tifaschi, *Azhar al Afkar fi Djawahir al Ahdjar*, 49.
3. Schumann, 140.
4. Al Tifaschi, 60.

CHAPTER 17

MAGNETITE (*MAGHNATIS*)

ONE should first note the similarity between the Arabic and the English terminologies used for this mineral, both being derived from the Greek. Although magnetite is not a gemstone nowadays, it was probably considered valuable in the thirteenth century due to its magnetic properties. Al Tifaschi says that this stone is found in the land between Hidjaz (Saudi Arabia) and Yemen, as well as in India, where huge stones suitable for use in buildings are cut. He tells that people in India have built a shrine to Buddha (which he described as a kind of statue they revere), whereby the walls, ceiling, and floor are all made of magnetite. The Buddha figure is made of iron and stands unsupported in the air in the very middle of the shrine, due to the strong attraction exerted on it from all sides by the magnetite. Magnetite is in fact strongly magnetic and can sometimes act as a magnet, this variety being called lodestone. According to al Tifaschi, the best variety was the one that was strongly magnetic, of a deep blue color similar to that of lapis lazuli. From this description, we can presume that he was referring to the compact granular iron-black magnetite that has a bluish iridescence.

Prices

An ounce of pure strongly magnetic magnetite sold for 1/4 dinar.[1]

Supposed medicinal benefits and superstitions

- If powdered and spread over a wound caused by rusted iron, it acts as an antidote against the poison.
- As a talisman or a bracelet, it helps combat rhumatism.

Notes

1. Al Tifaschi, *Azhar al Afkar al Djawahir al Ahdjar*, 63.

CHAPTER 18

EMERY (COMMON CORUNDUM) *(SUNBAZDAJ)*

EMERY is a finely granular impure black variety of corundum. It was found in a remote island off the coast of China. In fact, the island of Naxos in Greece supplied the best–and almost the only–emery in ancient and medieval times. As Greece is called "Yunan" in Arabic, it could have been confused for the "Yunan" province in China, and emery could in fact have come from Naxos and not China as reported by al Tifaschi, especially as he says: "It is mentioned that . . ." and not "emery comes from" as he usually does for most other stones.

Al Tifaschi compared emery to grains of sand of different sizes, the best being the large clear grains. This is a precise description of emery, which is a mixture of granular corundum, magnetite, and hematite and is used for abrasive and polishing purposes. Al Tifaschi said that this stone is very hard (hardness 3 on the Tifaschi scale and 9 on the Mohs scale) and very cool (degree of coolness of 2 on the Tifaschi scale). This is in accordance with modern classification, as emery is in fact impure corundum and has a hardness of 9 (right after diamond) on the Mohs scale. We see here another example of the precise scientific reasoning and observation of al Tifaschi, as he described emery as a stone separate from corundum, although he gave it the same degree of hardness on his scale. Al Tifaschi also pointed out that when powdered, emery is even more effective in grinding and polishing stones. This is indeed true, as when powdered, cutting edges are present in all directions.

Prices

An ounce sold for 1 *Dirhem.*[1]

Supposed medicinal benefits and superstitions

- It polishes and cleans teeth.
- It is used in medicines used to fight old age and skin flabbiness.

• If calcined and then powdered, it can heal eye sores and can help wounds heal.

Notes

1. Al Tifaschi, *Azhar al Afkar fi Djawahir al Ahdjar*, 64.

CHAPTER 19

MALACHITE *(DAHNADJ)*

MALACHITE is a basic copper carbonate. According to al Tifaschi, it was found in Iran and Arabia, in four different varieties, *afrandi, hindi, kirmani,* and *karaki,* the best being the banded emerald-green and light-green color, showing eyelike formations and readily accepting a good polish, this being the *afrandi* variety. The above is an exact description of good quality malachite, that shows the markings described by al Tifaschi. By the other three varieties, al Tifaschi could have meant azurmalachite, which is malachite intergrown with lazurite, and unbanded malachite.

Al Tifaschi noted that this stone improved in color when it was soaked in oil; however it could turn blackish (or bluish-black) if left to soak too long. Al Tifaschi also noted that this a rather soft stone (its hardness is only 3 1/2 - 4 on the Mohs scale), although it is sometimes used for making knife hilts, plates, and other such items. He gave this stone a degree of 4 on his heat scale (the highest number, together with ruby).

According to him, malachite is poisonous if drunk in powder form; its harmful effect and relative treatment being similar to that for iron rust.

Prices

1 *Mithqal* (24 carats) of the best quality was worth 2 gold *Mithqals.*[1]

Supposed medicinal benefits and superstitions

- If a small quantity in powder form (mixed with some liquid) is drunk, it prevents pregnancy. However, this constitutes a great risk to the person as mentioned earlier.
- If one kills seven flies with this stone, then powders it, mixes it with some water, and spreads it over bites from scorpions and snakes, it acts as an antidote against the poison.

Notes

1. Al Tifaschi, *Azhar al Afkar fi Djawahir al Ahdjar*, 67

CHAPTER 20

LAPIS LAZULI *(LAZUWARD* OR *LAZWARD)*

AS is the case today, the best lapis lazuli came from Afghanistan, at the border with Armenia, from a place known at the time of al Tifaschi as "Jestan" in the "Tukharestan" mountains. According to al Tifaschi, this stone is a soft, porous one, the best color being a deep sky blue neighboring on midnight blue.

Al Tifaschi's described a test to differentiate between natural lapis and its imitations, a method for the synthetic manufacture of imitation lapis lazuli, and the process used to "purify" lapis (that is to extract the ultramarine pigment). The test al Tifaschi described is none other than the modern "blowpipe flame test," which consists of heating the stone to be tested on a platinum wire and examining the flame on the outer part of the blowpipe.

Al Tifaschi described his experiment as follows: "If a piece of lapis lazuli is placed on smokeless coal, the resulting flame will be colored blue; while the color of the lapis lazuli stone remains unchanged."[1] Al Tifaschi specifically mentioned "smokeless coal," (probably charcoal, which is also used in modern tests), and coloration of the flame reveals the presence of specific elements in the stone.

As al Tifaschi noted that the resulting flame was blue, the possibilities are the following:[2]

Color produced	Substance causing it
Bluish-green	Phosphoric acid (phosphate)
Greenish-blue	Antimony
Azure blue	Chloride of copper-selenium
Whitish-blue	Arsenic
Violet	Potassium
Pale bluish	Tellerium
Blue	Lead

The presence of potassium imparts the violet/mauve color to the flame, but the presence of soda (in hauynite and sodalite, for example) is enough to mask the violet flame with a yellowish color. However, after heating, the color of the mineral remains violet on cooling.[3] Lapis lazuli is a rock composed mainly of hauynite, sodalite, noselite, and lazurite; therefore, it contains enough soda to mask the blue flame with a yellow color. How could al Tifaschi have noticed a blue flame? At any rate, his remark that the mineral remains blue after heating is correct, and the fact that flame tests were already used by the Arabs in the thirteenth century is remarkable.

Regarding imitations of lapis lazuli, these have been known to be manufactured in Egypt since a very long time ago. In his *Treatise on Stones*, Theophrastus mentions that the best manufactured *kyanos* (the old name for lapis lazuli) was made in Egypt.[4] This is known as "blue frit," a glass manufactured in ancient Egypt to imitate lapis lazuli.

Al Tifaschi described the following method that was used to manufacture material imitating lapis lazuli:[5] "A portion of red *zarnich* is taken, and another equal portion of yellow *zarnich*; (also) a quarter portion of vitriol from Kirman and a similar (quarter portion) one of clear quartz taken from pure sand. Each (ingredient) is powdered separately, then sieved and moistened with good vinegar. A clay container is taken and coated well with compact clay containing hair and manure, then it is left to dry. The ingredients are placed in it, moistened with just enough vinegar to reach the consistency of *sawiq*, then the (clay) container is closed with rags, and its aperture is coated with clay. The oven is lit with wood then with manure until the manure in it becomes as (thick as) the bone of the arm, and the clay container is placed in the oven, buried under the manure. The oven is closed with clay; lit at the bottom and left overnight. In the morning, the clay container is removed from the oven, and what is inside it is taken out. You will find small red stones, similar to ruby, round like pearls or elongated, the best one can see of stones." I do not understand how these red stones can resemble blue lapis lazuli, although al Tifaschi repeatedly tells us that he described his experiment to show the reader how manufactured lapis is made.[6] Nevertheless, I have pointed out this passage to show that artificial, nonglass stones were made as early as the year 1240. As a matter of fact, al Biruni had already described this very same experiment in his book[7] (see later in the "imitations and synthesis of gemstones" chapter), saying that he had taken this recipe from "ancient books." Whether these ancient books were Western or Arab, he did not explain.

The "purification" of lapis lazuli by washing it and removing all impurities was apparently a noble profession exercised by many people. This purification process was none other than extracting the ultramarine pigment from the stone. From the intricate process described by al Tifaschi, we can deduce that the ultramarine pigment that was extracted at that time was of a much better quality than the modern one, which is obtained by powdering lapis lazuli without removing its impurities.

Prices

A good "purified" lapis stone suitable for use in finger-rings was worth approximately 3 *Dirhems*. The price of 1 ounce of pure lapis was worth up to 1 dinar, while that of the rough unworked stone was worth 2/3 of the above price.[8]

Supposed medicinal benefits and superstitions

- If added to a collyrium, it is beneficial for the eyes.
- Applied in powder form, it can speed growth of eyelashes.
- It helps cure melancholy (it is interesting to note here that al Tifaschi used the exact pronounciation of the word in Greek, *Melankholia).*
- It helps overcome some forms of fever.
- Drinking a little lapis mixed with some honey can cure liver disorders.
- Mixed with vinegar and spread over the affected area, it helps cure leprosy spots.
- Worn as a talisman, it protects children from fear and anxieties.
- Mixed with any hair lotion, it improves the texture of the hair and makes it curly.

Notes

1. Al Tifaschi, *Azhar al Afkar fi Djawahir al Ahdjar*, 67.
2. Dana, *A Textbook of Mineralogy*, 364.
3. Ibid., 590.
4. Caley and Richards, *Theophrastus on Stones*, 184.
5. Al Tifaschi, 68.
6. Ibid., 69.
7. Al Biruni, *Al Djamahir fi Ma'arifat al Djawahir*, 227.
8. Al Tifaschi, 175.

CHAPTER 21

CORAL (*MARDJAN*)

CORAL is composed mostly of calcium carbonate with some magnesia and organic substances. It grows on reefs and atolls in branchlike formations. The best coral is the "noble coral," of a deep red color. Other shades include white, pink, and orange-pink. Another type of coral is made of a hornlike substance and comes in black, golden, brown, and blue.

The Mediterranean waters (North African and Italian waters) and the Red Sea have always supplied the best coral. Nowadays, coral is also found in Malaysia, Japan, and Australia. According to al Tifaschi, Alexandria was the greatest center of the coral industry, the place where it was polished, engraved, carved, and pierced for use in jewelry.[1] Nowadays, the center of the coral industry is in Italy at Torre Del Greco, near Naples.

The modern description of coral is that it is "a scaffolding upon the surface of which the boneless animals live as a colony."[2] Al Tifaschi described coral as "midway between the two worlds of plants and inorganic substances. It resembles minerals in petrification and resembles plants in its treelike formations."[3]

Theophrastus could not decide whether coral should be classified as a stone or as a plant. Pliny could not decide whether it should be classified as a plant or as an animal.[4] In the first edition of his book, J. Hill decided that coral is a plant, stating "*j'ai des preuves qui feront voir que les conséquences qu'on en peut tirer, favoriseront toujours l'ancienne opinion que le corail est un végétal* "[5] (I have proofs that will show that the consequences one can deduce will always favor the old opinion that coral is a plant). But he later re-classified it as a mineral in the second edition,[6] as he did not believe that coral was some kind of fossil, nor animal, nor vegetable. In *Introduction to the Study of Recent Corals,*[7] Hickson said that the animal nature of coral was not understood until the middle of the eighteenth century. This seems to have been the case, as al Tifaschi also considered coral as a species midway between the plant and the mineral kingdoms.

Al Tifaschi mentions a place called Marsa al Kharaz in the sea of

Tunis and the "Sea of Europe" (probably meaning French and Italian waters) as yielding the best coral. The best kind, according to him, is the bright red, free of black spots, the whitish shade being the least valuable. No mention of the horn types of coral was made.

When large pieces of good red color were found, they were reserved and made into special items for the king of Tunisia, such as inkstands, and dagger hilts, etc.

Al Tifaschi warned that vinegar can turn red coral into white and can even dissolve it if one leaves coral to soak in it for a long time. On the other hand, oil and grease improve the color of coral and give it sheen and lustre. He gave coral a degree of hardness of 3 and degree of coolness of 1 on his scales.

Al Tifaschi described with great precision a method for etching on coral: "If a person wants to write something he likes (on coral), he covers the whole stone or the whole ring with wax; then he engraves with the tip of a needle what he likes until the wax is removed from the place of the engraving only. Then he soaks the stone in strong vinegar for one day and one night, or two days and two nights, then he removes it and removes the wax from it; he finds the place where the etching is engraved, eaten away by the vinegar, while the rest of the stone remained as it was, unchanged."[8] It is interesting to note that this is the exact method that M. Bauer described in his book *Precious Stones,* volume 1, when he said: "The polished surface of the stone to be etched is covered with a thin coating of wax, upon which the outline of the device is drawn with some sharp instrument; the surface of the stone along this outline is therefore laid bare while other parts are protected by the wax. The stone is then placed in liquid or gaseous hydrofluoric acid which eats away the surface of the stone where it is not protected by a layer of wax . . . .After removing the wax and cleaning the stone, the device . . . sketched upon the wax will appear as if cut in the stone."[9] Bauer was acquainted with al Tifaschi's writings, as is clear when he referred to him on page 257 of his book (where he called him Teifaschius), regarding prices of diamond. His description of the above etching method was obviously inspired by the method described by al Tifaschi, which could be the first account of etching in the history of gemology.

Coral was ground with water on the whetstone and polished with emery on a marble slab (or wheel?). A steel point was used for drilling coral.

Prices

Large pieces commanded higher prices than smaller ones. Prices also differed according to rarity of the coral. In North Africa, where coral is found, the price of 1 North African *Ratl* (16 ounces) was worth 5-7 *Maghribi* dinars. However, in Alexandria, prices were double the ones mentioned above and increased according to availability of the material.[10]

Supposed medicinal benefits and superstitions

- If worn as a talisman by children, it protects them against the evil eye and evil spirits.
- It is beneficial for the eyes when added to a collyrium; it strengthens them and removes scars and other dirt that might accumulate in them.
- When drunk (after being powdered), it is effective against heart palpitations and other heart diseases; this is why it is incorporated in medicines used to prevent blood clots.
- It fortifies the blood and is effective against dysurea.
- If worn over the stomach, it is effective against all types of stomach pains.
- It can stop hemorrhage from any part of the body. For this purpose, a portion of coral is powdered after being calcined, then it is mixed with half a portion of gum Arabic. The mixture is bound with eggwhite and the remedy drunk with cold water.
- Calcined coral is also effective against tooth decay and gum diseases.

Notes

1. Al Tifaschi, *Azhar al Afkar fi Djawahir al Ahdjar*, 74.
2. Webster, *Gems: Their Sources, Descriptions, and Identification*, 562.
3. Al Tifaschi, 70.
4. Caley and Richards, *Theophrastus on Stones*, commentary, 141.
5. Hill, *Theophrastus History of Stones*, 1st ed., 140 (in French).
6. Hill, *Theophrastus History of Stones*, 2nd ed., 168-169 (in English).
7. Hickson, *An Introduction to the Study of Recent Corals*, 11-14.
8. Al Tifaschi, 73.
9. Bauer, *Precious Stones*, vol. 1, 87.
10. Al Tifaschi, 74.

CHAPTER 22

JET *(SABADJ)*

JET is black fossilized wood, allied to lignite or brown coal. It is formed when decomposed wood is compressed over millions of years, allowing no cell cavities to be present. It is interesting to note that the Spanish word for jet, namely *"Azabache"* comes from the Arabic "*As Sabadj*." As the Spanish language has no "j" pronounciation, the "j" was replaced by "che" and the resulting word became "*Azabache.*"

Al Tifaschi stated that lead enters in the composition of this stone, which was found mainly in India and Persia. The best variety was the Indian one, which was opaque and of a deep black color. Jet can be highly polished, and thus was used for mirrors. Al Tifaschi also said that this stone is very soft and easily breakable, giving it a hardness of 1 and a degree of coolness of 1. Simulants of jet include cannel coal among others, which has a black streak and can thus be differentiated from jet, which displays a brown streak.[1] However, as India and Persia are the sources mentioned by al Tifaschi (and not England and Scotland, where cannel coal comes from), the word *Sabadj* referred probably exclusively to jet.

Prices

A bead weighing 1 *Mithqal* (24 carats) was worth 1/2 *Dirhem.*[2]

Supposed medicinal benefits and superstitions

- If powdered and added to a collyrium, it is beneficial for the eyes.
- It can improve vision if one stares at it.
- As a talisman, it can protect fair-skinned people (or people wearing white clothes) against the evil eye.

Notes

1. Muller, *Jet*, 120.
2. Al Tifaschi, *Azhar al Afkar fi Djawahir al Ahdjar*, 75.

CHAPTER 23

AMETHYST *(DJIMASHTT)*

AMETHYST is a crystalline variety of the quartz family of a lovely mauve to deep purple color. Before the rich South American sources were discovered in the nineteenth century, amethyst was considered to be a rare and valuable gemstone. Al Tifaschi says that this stone was highly esteemed and was used for decorating swords and other weapons, as it was supposed to instill courage in the person in times of battle.

The color of amethyst puzzled gemologists for centuries. As R. Webster explains:

> The cause of colour in amethyst had long been a subject for conjecture and a number of theories have been put forward. A suggestion at one time was that manganese was responsible, the idea being derived from the fact that specimens of glass containing manganese assumed a similar tint. Thanks to recent researches and to the successful synthesis of amethyst by hydrothermal methods, it is now certain that *the amethyst hue is due to* radiation induced colour centres associated with *an iron impurity.*[1]

Until the twentieth century, the true cause of the color of amethyst was still unknown in the West until sophisticated electronic equipment was able to detect radiation-induced color centers, revealing that iron was responsible for its color. What is truly amazing is that al Tifaschi stated in the very first line of the chapter on amethyst that iron forms the basis of this stone's composition.[2] How could this thirteenth-century scholar have known what twentieth-century scientists only discovered recently with electronic equipment?

Al Tifaschi mentioned four shades of amethyst, ranging from light pinkish mauve to deep purple, saying that the best color is deep violet (an equal mixture of strong red and blue). Amethyst was cut (ground) on a "bed of lead," probably meaning that a lead sheet, applied to the traditional Indian lap wheel, was used to polish the stone. After grinding, amethyst was polished on wood of the agigantea tree.

The only known location for amethyst at that time was Saudi Arabia, near Mecca, in a village called Safra located between Medina and Yanbu.

Prices

A stone of 1/2 *Dirhem* was worth 1 *Dirhem*[3].

Supposed medicinal benefits and superstitions

- Amethyst instills courage in the wearer in times of battle.
- Worn as a talisman, it protects against gout.
- Placed under a pillow, it protects the dreamer against nightmares.
- Quoting the famous tenth-century Arab physician al Razi, al Tifaschi said that whoever makes a drinking glass of amethyst and drinks in it as much wine as he likes, he will not get drunk. Thus, it is to the Arabs that we owe the belief that amethyst could protect against the ill effects of excessive alcohol, which is in line with the origin of the name amethyst, *a-methystos* in Greek, meaning not drunk. Theophrastus had explained that the name was given to the stone because its color resembles that of wine, while, on the contrary, Pliny said that the name was given because the color of the stone was near that, but not quite, that of wine.

Notes

1. Webster, *Gems: Their Sources, Descriptions, and Identification*, 209.
2. Al Tifaschi, *Azhar al Afkar fi Djawahir al Ahdjar*, 75.
3. Ibid., 76.

CHAPTER 24

HEMATITE *(KHUMAHAN)*

THE name probably came from the Greek (*Hema* means blood) as when it is cut, the coolant is colored red, and the stone has a red streak powder resembling dried blood. In Germany, hematite is called blood stone. But heliotrope (also called plasma), a dark green stone with red spots of jasper belonging to the cryptocrystalline chalcedony group, is also called bloodstone in other countries. Al Tifaschi mentions that this stone is also called *al Sirf*, which is Arabic for a red dye and also the name of a tree called the "blood tree." However, there could also be another explanation for the name *Sirf*, for later on in the text, he says that drinking a little powdered hematite can help cure hangovers resulting from drinking pure alcohol (called *al Sharab al Sirf* in Arabic). The above controversy is one of many examples where the same nomenclature given to different stones has lead to confusion. But it is interesting to note that hematite is associated with blood in many countries, probably due to its streak powder, which must have been striking in old days as the stone is of black color.

An important point al Tifaschi made when describing this stone is that hematite is composed of iron, hematite indeed being a black iron oxide. The best and only source for hematite during al Tifaschi's time was Karak, in Jordan.

Prices

1 *Ratl* (12 oz.) sold for 3 *Dirhems.*[1]

Supposed medicinal benefits and superstitions

Drinking a little powdered hematite can cure hangovers resulting from drinking pure alcohol.

Notes

1. Al Tifaschi, *Azhar al Afkar fi Djawahir al Ahdjar*, 77.

CHAPTER 25

JADE (NEPHRITE) *(YASHM)*

THE name "jade" refers to two minerals, namely nephrite and jadeite. By "*Yashm*," al Tifaschi meant nephrite, a calcium magnesium iron silicate forming a compact mass of fibrous crystals. This structure gives it an exceptional hardness, and this is the reason nephrite was used in weapons beginning in prehistoric times. Nephrite occurs in many colors, but the white varieties called "sodden snow" or "mutton fat" and the dark green shades are the most common and well known and come from Turkestan. "Buried jade" is jade that has remained buried since prehistoric times and has altered in color to pinkish, yellowish, grayish, or brownish. This jade is found in China.

He stated that two kinds of nephrite were commonly used by people: a yellow variety and another white one, the latter being "man-made." This white variety, probably porcelain, was supposedly manufactured in China and exported all over the world. It was used mostly for bowls of all sizes and shapes, as well as for other utensils. Al Tifaschi mentioned that he saw many items of this white variety being sold in Cairo as genuine nephrite, while in fact they were not. He believed that the white variety was a manufactured one, as he himself had manufactured "white nephrite" and given it to an emir who collected items made of this stone. As the emir refused to believe al Tifaschi, they agreed on items of specific shapes to be manufactured, which al Tifaschi did, thus proving to the emir that they were indeed man-made.[1]

Al Tifaschi mentioned that nephrite was found in Kashgar, where it was sent all over the world. Al Biruni had stated that nephrite was bluish and came from a valley in Khutan (China). It is indeed interesting to note that these are the exact two locations where the most important deposits are still found nowadays.

Prices

Bowls made from nephrite were brought from China to be sold in Cairo. A bowl of 1 to 2 1/2 *Mna* (1 *Mna* was equal to approximately 794 g) sold for 5 Egyptian dinars, and large bowls brought as much as 50 dinars. A stone suitable for a finger-ring was worth 4 *Dirhems*.[2]

Supposed medicinal benefits and superstitions

• Lightning never strikes the place where nephrite is found. Al Tifaschi mentions that some "foreign" people (in this context he meant Eastern or Far Eastern people) told him that they saw towers built in their countries in a place where lightning often strikes. On top of these towers were light-houses where nephrite stones were hung well in sight. One could see lightning falling everywhere around these towers but never hitting them, due to the presence of these stones.

• Worn in a ring, this stone can dispel nightmares.

Notes

1. Al Tifaschi, *Azhar al Afkar fi Djawahir al Ahdjar*, 77.
2. Ibid., 78.

CHAPTER 26

JASPER *(YASSB)*

JASPER is a microcrystalline variety of quartz. Some authors are of the opinion that jasper was known as *haimatitis* by Theophrastus and *haematites* by Pliny. However, the name "hematite" nowadays refers to a black iron oxide with a red streak powder resembling dried blood.

Very little was written on this stone. Al Tifaschi simply mentions that it is found in two colors, white and green, green being the best variety. Today also, green varieties of jasper are considered the most valuable. Al Tifaschi remarked that this stone can easily be dyed; it was often dyed blue, and some jewelers even dyed it green to resemble emerald. As a matter of fact, "Swiss lapis" or "German lapis" are modern imitations of lapis lazuli made of jasper stained blue with potassium ferrocyanide and ferrous sulphate.[1]

Prices

A stone suitable for a finger ring, weighing 1 1/2 *Dirhems* was worth 1 *Naqara Dirhem.*[2]

Supposed medicinal benefits and superstitions

One medicinal benefit only was mentioned by al Tifaschi, for which he quoted another Arab scholar: that jasper was beneficial for stomach aches when worn externally.

Notes

1. Wesbter, *Gems: Their Sources Descriptions, and Identification,* 253.
2. Al Tifaschi, *Azhar al Afkar fi Djawahir al Ahdjar*, 78.

CHAPTER 27

ROCK CRYSTAL *(BILLAWR* or *BALLUR)*

BILLAWR or *Ballur* (same word) refers exclusively to rock crystal in Arabic. Rock crystal is a crystalline variety of quartz and consists of silica. The first crystals were discovered near Mount Olympus, Greece, and were called *Krystallos*, or frozen ice, hence the origin of the name.

During al Tifaschi's time, it was found in Barriyet al 'Arab (Arabian desert), near Hidjaz in Saudi Arabia, which provided the best quality, as well as in *Bilad al Afrandj* (Europe). It was also found in China and Armenia, the latter displaying a yellowish tinge typical of this location. Rock crystal was also abundant in Morocco and was used to decorate walls and floors of palaces. Al Tifaschi tells of two huge mountains of rock crystal located between Kashgar and Ghaznah (between Chinese Turkestan and Afghanistan) that were worked only at night, as it was impossible to work them by daylight due to the strong, blinding reflections and refractions from the surface of the stones.[1]

Another reference made by this author is of some importance, as al Tifaschi mentions that one of the characteristics of rock crystal is that if a piece of cloth is placed beneath the stone so as to receive direct sunlight through this stone, this cloth will burn. This is an indirect reference to the use of rock crystal in magnifying lenses.[2]

Al Tifaschi also described a jug carved somewhere in Europe from pure rock crystal in the shape of a rooster; when liquor was poured in it, it appeared even in the claws of the rooster due to the intricacy of the carving. Many large objects were carved from rock crystal as early as Roman times, including dishes, bowls, and drinking vessels.

Mention was also made of a rich European merchant of the time who had given one of the kings of Morocco a canopy with a dome of rock crystal, able to seat four persons. The king of Ghaznah had in his palace four water containers, each able to hold three skins of water, held by supports each weighing 3 to 4 *Qintars* (approximately 135 to 180 kgs), all of rock crystal.

The best rock crystal was described as being the clearest, most transparent, colorless stone, free of cleavage or incipient cracks.

Prices

Prices differed according to the size and quality of the material. The wholesale price of 1 *Ratl* (12 oz) of best quality material was 10 dinars.[3]

Supposed medicinal benefits and superstitions

The only benefit of rock crystal was its ability to dispel nighmares if worn as a talisman.

Notes

1. Al Tifaschi, *Azhar al Afkar fi Djawahir al Ahdjar*, 81.
2. Ibid., 81.
3. Ibid., 81.

CHAPTER 28

TALC (*TALQ*)

TALC is a hydrous magnesium silicate, consisting mainly of the mineral talc. The word *Talq* is believed by Arab lexicographers to be the original Arabic name of the mineral, from which the West derived talc and *talco*. Colors are white, silvery-white, greenish-white, yellowish, and light gray or brownish.

The only source for talc in the thirteenth century was Cyprus, where it was found in "silvery" or "golden" colors. The "silvery" shade referred to the pure white mineral and the "golden" one to the white mineral with a yellowish tinge. Al Tifaschi mentioned that talc is light (he said it was actually formed like dew drops), but somehow very hard, as it is infusible and has great resistance to heat and fire. Also, it can be cut into small slithers, rather than broken into pieces. That talc gained a high degree of hardness after heating was known since Theophrastus's times, as he mentioned that dishes and utensils were made of talc (our modern soapstone).[1] As a matter of fact, talc has great resistance to heat and fire and is flexible. It can be bent without breaking, remaining bent after the bending force is removed.[2] The only way to powder talc, according to al Tifaschi, is to put some talc stones with other small pebbles in a bag made of some rough material, and place the bag in hot water where fava beans have been cooked, shaking it constantly. The talc can thus be powdered and when the bag is lifted from the water and dried, a fine powder is obtained.[3]

Although modern textbooks mention that talc is also insoluble, al Tifaschi described a method which was used to dissolve talc into a solution similar to "agitated water" (meaning soft, gelatin-like).* The method was as follows: "Sandarac resin is taken and pounded until it becomes fine; then it is placed in a crucible and tinkal and natron are poured on it and it is melted until it becomes like water."[4] This solution could apparently dissolve talc, iron, and other minerals as well.

*This paragraph is not in the Bibliotheque Nationale's version. It was taken from the Cairo one.

Al Tifaschi described a method for using talc in housepaint to give a pearly luster to the house.[5] Nowadays also, powdered talc is used in the paint industry. Talc was also used to render bodies incombustible, and one way to preserve ships from the action of fire was to coat them with a layer of the above-mentioned solution, to which resinous gum, red clay, and alum were added.

Prices

One ounce of the best quality sold for 1 *Dirhem.*[6]

Notes

1. Caley and Richards, *Theophrastus on Stones*, 146.
2. Simon and Schuster's *Guide to rocks and minerals*, 224.
3. Al Tifaschi, *Azhar al Afkar fi Djawahir al Ahdjar*, 82.
4. Hassan and Khafaga, commentary, 206.
5. Ibid., 206.
6. Ibid., 206.

PART IV
ANALYSIS OF GEMOLOGY IN AL TIFASCHI

CHAPTER 29

Fashioning, Grinding, and Polishing of Gemstones

THE art of cutting and polishing gemstones is poorly documented and remained secret for many years. In Europe, and until somewhere in the fourteenth century, cutting of gemstones was limited to polishing the natural crystal faces of the stones (or the cleavage planes), cutting the stone into *cabochon* style, or drilling the stone to make a hole suitable for attachment in a pendant, an amulet, or a talisman.

In his 1638 trip to India, J. P. Tavernier noted that lapidaries in India valued stones on the basis of quality and weight only. Therefore, the only polishing they did was to remove or hide defects in the stone.[1] Many books on diamond also attribute the discovery of polishing and faceting to Berghem of Bruges in 1476.[2] However, in the Arab and Islamic world, and as evidenced by the writings of al Biruni and al Tifaschi, stones were cut by experienced, specialized workers into specific forms and shapes at least from the times of al Kindi, that is, the ninth century.

Al Kindi had reported on a block of sapphire that was cut into the shape of a slave girl sitting and clutching her knees between her interlocking fingers. Furthermore, al Biruni mentioned that "specialists" were often called to cut a particularly precious and important stone. This "specialist" was called *Kharrat* (plural *Kharratun*), which means "cutter(s)." In another passage in the ruby chapter, al Biruni mentions that "jewelers state today that a ruby stone of the *rummani* kind, with a flat table, *of a symmetrical rectangular or square shape, if such is the shape desired,* then the *tassouj* of this stone starts at 5 dinars, doubling in value with the doubling in weight."[3]

The above sentence needs careful review and explanation to grasp all the information contained in it: The *tassouj* is a unit of weight for precious stones, and al Biruni explained that it was equal to 1/3 of 1/8th of a *Mithqal.* The *Mithqal* being 24 carats, therefore, the *tassouj* was the word used to describe 1 carat. Al Biruni's statement "with a flat table, of a symmetrical rectangular or square shape" implies that, apart from the old *cabochon* cut, stones were also cut into rudimentary forms of *baton, baguette,* and square cuts. However, the most important part of the sentence is: "If such is the shape desired." In Arabic, the sentence transliterates as follows: *Iza kana huwa'l mukhtar min Ashkalihi,* and the word *Ashkalihi* which means "shapes," "forms," or "appearances," could be more appropriately translated here as "cut." Therefore, it is clear from the above, that different cuts were used for gemstones even before al Biruni's time. Whether facets were used or not is not known, but truncated octahedron diamonds were available in the Arab world, having been rejected by the Indians, who cared only for the perfect, unbroken, well-formed stones. The *Kharratun,* or specialists in cutting gemstones, were sent for from one country to the other when a particularly valuable specimen needed shaping, as al Biruni explains under the turquoise chapter: Apparently, the Emir Noah ibn Mansur (A.D. tenth century) had a bowl of turquoise that could contain 3 *Ratls* (36 oz.) of liquid. A special cutter was sent for from Iraq (presumably to alter the shape of the stone), but unluckily the stone broke into his hands while he was fashioning it.[4]

As for al Tifaschi, he mentioned that stones were "improved" or "worked" in several instances, and he repeatedly mentioned when describing many stones (especially ruby and emerald) that "the stone loses a lot of its weight after grinding *(mahakk),* polishing *(djala')* and working it to give it the final look *(handama).*[5] A gemstone could apparently often be bought very cheap when rough and resold at exorbitant prices after "adjusting." The exact translation of the Arabic word he used, namely *handama,* is "fashioning it to make it better, to obtain the final look." He gave the example of an emerald that was bought as a rough crystal for 13 *Dirhems** and resold after "grinding, polishing, and working and finalizing" to a rich ruler for 30,000 *Dirhems.*[6] The difference is exorbitant indeed, and we can only assume that something must have been done to that stone to make it so attractive!!

Frequent mention is also made of the hollowing of some stones to make their color more attractive, particularly so in the case of almandine garnet. This procedure is still being carried out all over the world today, and red stones (especially almandines) that were treated this way were known as "carbuncles" in Victorian times.

Regarding polishing, al Tifaschi explains that all stones can be polished on wood from the agigantea tree[7] except ruby (and other corundums). Ruby was polished by placing it on a copper lap and using burnt and

* The price in question was mentioned in the Cairo version only.

calcined agate mixed with a little water into a thick paste. Apparently, no other method of polishing could give ruby such a strong luster as that obtained with agate. Powdered emery mixed with a little water (water was added to act both as a coolant and as a binding agent for the powdered polishing agent) was used for grinding amethyst and emerald, then polishing was carried out on a wooden lap made of the same agigantea tree. Even today, copper laps are used for grinding, and wooden ones for polishing gemstones other than diamond, especially in places where more advanced diamond-charged laps are not available.

Notes

1. Tolansky, *The History and Use of Diamond*, 70.
2. Ibid., 70.
3. Al Biruni, *Al Djamahir fi Ma 'arifat al Djawahir*, 49.
4. Ibid., 170.
5. Al Tifaschi, *Azhar al Afkar fi Djawahir al Ahdjar*, 89.
6. Ibid., 89.
7. Ibid., 17.

CHAPTER 30

Imitation and Synthesis of Gemstones

THE medieval Arabs did not use the words "genuine" or "synthetic" to differentiate between stones. Rather, they called genuine stones *makhluq* (meaning "natural" or "born"), and imitation and synthetic ones *masnu'u, mudallas,* or *maghschusch* (meaning "man-made," "fake" or "cheated").

The first synthetic crystals (ruby) were made by a French scientist in 1837. In 1877, another French scientist, E. Frémy, published a paper on the manufacture of synthetic gemstones, and one of his assistants, Auguste Verneuil, perfected the method that nowadays carries his name and is widely used for the commercial production of synthetic corundum and other stones. It is only in the twentieth century that the production of synthetic gemstones has become widely practiced, and several methods are available for this purpose, the two basic procedures being:

1. Crystallization from a melt: The powdered chemical components are mixed and melted, then cooled to allow crystallization. A method called flame fusion is used to produce mainly corundum and spinel; another method called zone melting is used to produce corundum and alexandrite; quartz is manufactured by the crystal pulling method and cubic zirconia by skull melting

2. Crystallization from a solution: The chemical components are dissolved in a liquid in the hottest part of the container, and crystallization takes place in a cooler part: The hydrothermal method is used to produce emerald and quartz and the flux-melt growth to produce corundum and emerald.

Al Tifaschi's method of synthesis, which has already been described in the chapter on lapis lazuli, proves that artificial stones were manufactured over 750 years ago. As a matter of fact, it was al Biruni who first described this method in his book. He explained that a stone called *Azhrak* (could it have been one of the famous anthrax stones of Theophrastus?) was a very valuable and old red glass that was manufactured by the ancient Greeks, the descendants of Alexander. When pieces of this rare material were found, they sold for 1,000 dinars a piece during the times of al Kindi, as they were extremely rare and could not be manufactured again, the process having been lost. Al Kindi explained that people had tried to manufacture this type of glass for the *Khalifa* al Mutawakkil (ninth century) but that only a lighter-colored variety was obtained.[1] Apparently *Darnuk* was another artificial stone of red color mixed with orange that was also manufactured by the ancient Greeks and was very valuable and rare.

From al Biruni's and al Tifaschi's description of the above method of synthesis, we note that it resembles two modern methods of gemstone synthesis: The hydrothermal method, whereby raw materials are enclosed in a partially filled sealed container of alkaline water, which is heated; the container, being hermetically sealed, the hot dense steam dissolves the materials, and when the container is cooled, crystallization occurs. The other method is the flux-melt growth whereby a mixture of chemicals are heated in a crucible, which produces the solvent for the chemicals required to form the crystals. Al Tifaschi's method is a sort of hydrothermal method, whereby vinegar is used instead of alkaline water and where no seed plates are present; thus, small crystals were obtained (according to al Tifaschi's own description).

In his book, *Gems: Their Sources, Descriptions, and Identification,* R. Webster states that "ruby, the red variety of corundum, is a stone whose value is sufficiently high to warrant consideration of its synthesis. The earliest recorded experiments in this direction were those of Marc A. A. Gaudin, who produced a few tiny flakes of crystallized alumina by heating a previously ignited mixture of alum and a little potassium chromate covered with lamp black in a clay crucible. The experiment had no commercial significance, for although the globules were quite clear when molten, they became opaque on cooling. Although many experiments were tried, in his last published notes, in 1869, Gaudin virtually admitted defeat. However, this was the beginning of the history of the synthetic stones now used in jewelery."[2] In fact, the history of synthetic stones had begun much earlier, that is, almost 900 years earlier with al Biruni's first report on how to synthesize stones. As he himself was quoting another earlier author on this subject (whether Greek, Persian, or Arab he did not specify) it is unknown to exactly how far back the history of synthetic gemstones dates. It is also interesting here to note that Clément Mullet, a compatriot of Gaudin, published in 1868 under the title "Essai sur la Minéralogie Arabe" a report and a translation of some passages of al Tifaschi's book. One can easily note the similarity of Gaudin's method with the one described by al Biruni and al Tifaschi, whereby in both:

- A clay crucible was used;
- the experiments took place in a reducing atmosphere (meaning without the presence of oxygen, the crucible being hermetically sealed);
- red globules were obtained.

Thus, it could well be that Gaudin's method was in fact inspired by al Tifaschi's own method, as reported by Clément Mullet in 1868.[3]

From al Tifaschi's writings, we note that imitations of many stones were being manufactured routinely in the thirteenth century, including imitation pearls, nephrite, lapis lazuli, ruby, and other corundums. Even coral was imitated by planting wood sticks resembling coral spikes on the banks of rivers where deposits and mud residues were known to petrify (near Ghaznah in Afghanistan), then were removed, painted red, and sold as genuine corals.[4]

Notes

1. Al Biruni, *Al Djamahir fi Ma'arifat al Djawahir*, 227.
2. Webster, *Gems: Their Sources, Descriptions, and Identification*, 381.
3. Mullet, "Essai sur la Minéralogie Arabe,"505.
4. Webster, 193.

BIBLIOGRAPHY

'Abd Rab el Nabi, M.Y. "The Arabs and Geology" (in Arabic). *Al Faisal* (Saudi Arabia)(December 1991/1992).

Al Biruni, A. R. *Al Djamahir fi Ma'arifat al Djawahir*. Edited by Salem al Karankawi. Hyderabad: The Ottoman Organization of Da'irat al Ma'arif, 1939.

Al Bustani, B. *Muheet al Muheet*. Beirut: Librairie du Liban, 1977.

Al Jaser, H. "Old Mining Sites in the Arab countries" (in Arabic). *Al 'Arab* (Saudi Arabia)(August 1968).

Al Khatib, M. "Jewelry and Precious Stones" (in Arabic). *Al Faisal* (Saudi Arabia)(February/March 1986).

Al Qabadjaqi, B. *Kinz al Tudjar fi Ma'arifat al Ahdj*ar (The secrets of merchants with regards to the knowledge of stones). Photocopy of original Arabic manuscript, 1282. Bibliothèque Nationale, Paris, France.

Al Qalqashandi, A. *Subh al A'sha fi Sina'at al Insha*. Cairo, Egypt: Ministry of Education, 1963.

Al Tifaschi, Ahmad ibn Yusuf. *Azhar al Afkar fi Djawahir al Ahdjar*. Edited and with an introduction and notes by Drs. M. Y. Hassan, and M. B. Khafaga. Cairo: The Egyptian General Organisation for Books, 1977.

Anderson, B. W. *Gem Testing*. 9th ed. England: Butterworths and Co., 1988.

Aristotle. *Kitab Al Ahdjar li Aristatalis*, (Book on Stones). Translated by Luqa ibn Serafion. Photocopy of original Arabic manuscript, estimated to be from the middle of the 9th century. Bibliothèque Nationale, Paris, France.

Ball, S. H. "Historical Notes on Gem Mining." *Economic Geology*, vol. 26, no. 7 (1931).

Bamborough, P. *Treasures of Islam.* England: Blandford Press, 1976.

Bauer, J., and V. A. Bouska. *A Guide in Color to Precious and Semi-Precious Stones*. London: Octopus Books, 1983.

Bauer, M. *Precious Stones*, 2 vols. New York: Dover Publications, 1968.

Bin Salam, A. R. "The Importance of Precious Stones" (in Arabic). *Al Manhal* (Saudi Arabia)(November/December 1975).

Bruton, E. *Diamonds*. 2nd ed. England: N.A.G. Press Ltd., 1981.

Caley, E. R., and J. F. C. Richards. *Theophrastus on Stones*. Introduction, Greek text, English translation and commentary. Columbus, Ohio: The Ohio State University, 1956.

Cowan, J. M. *The Hans Wehr Dictionary of Modern Written Arabic*. 3rd ed. New York: Spoken Languages Services Inc., 1976.

Dana, E. S. *A Textbook of Mineralogy with an Extended Treatise on Crystallography and Physical Mineralogy*. 4th ed. Revised and enlarged by W. E. Ford. New York: John Wiley & Sons, 1932.

De Barrera, Madame. *Gems and Jewels, Their History, Geography, Chemistry, from the Earliest Ages down to the Present Time*. London: Richard Bentley, 1860.

Dietrich, R. V. *The Tourmaline Group*. New York: Van Nostrand Reinhold,1985.

Dozy, R. *Supplément aux Dictionnaires Arabes*. Leyde: E.J. Brill, 1881.

Encyclopedia of Islam, vols. 1, 2 and supplements 3, 4. Leiden: E.J. Brill, 1981.

Erythraspis, J. I. *Arab Art as Seen through the Monuments of Cairo from the 7th Century to the 18th.* Paris: Le Sycomore; London: Al Saqi Books, 1983.

Fraquet, H. *Amber.* Butterworths Gem Books, England: Butterworths & Co., 1987.

Gubelin, E., and J. Koivula. *Photoatlas of Inclusions in Gemstones.* Zurich: ABC Editions, 1986.

Hickson, S. J. *An Introduction to the Study of Recent Corals.* Manchester: University Press of Manchester, Biological Series no. 4, 1924.

Hill, J. *Theophrastus's History of Stones with an English Version and Notes.* 2nd ed. London: n.p., 1774.

Hill, M. *Traité des Pierres Précieuses de Théophraste. Traduit du Grec, avec des Notes Physiques et Critiques, traduites de l'Anglais.* Paris: Jean-Thomas Herrissant, 1754.

Hitti, P. K. *History of the Arabs.* 10th ed. New York: St. Martins Press, 1991.

Hunger, R. *The Magic of Amber.* London: N.A.G. Press Ltd. (in conjunction with Northwood Publications Ltd.) and Rosa Hunger, 1977.

Ibn abi Usaybi'a. *Tabaqat al Atibba'a.* Beirut: Dar Maktabat al Hayat, 1965.

Ibn al Akfani. *Nukhab al Zakha'ir fi Ahwal al Djawahir.* Edited by Father Anastase-Marie de Ste. Elie, Egypt, 1939.

Ibn al Nadim. *Kitab Al Fihrist.* (10th century). Edited by G. Flugel, Reprint. Beirut: Khayat's, 1964 .

Ibn Hunayn al Hakim, I. *Kitab Khawass al Ahdjar,* (Talismanic and magical properties of stones). Photocopy of original Arabic manuscript. n.d. Bibliothèque Nationale, Paris, France.

Ibn Mohammad al Katib, U. *Kitab Khawass al Ahdjar,* (Talismanic and magical properties of stones). Photocopy of original Arabic manuscript. n.d. Bibliothèque Nationale, Paris, France.

Kikano, A.B. *Table de Concordance des Années Hégiriennes et Chrétiennes*. n.p. Beirut, 1966.

King, C. W. *The Natural History of Gems or Semi-precious Stones*. London: Bell and Daldy, 1870.

Lentzen, G. *Diamond and Diamond Grading*. Translated by P. B. Lapworth. England: Butterworths & Co., 1983.

Mottana, A., R. Crespi and G. Liborio. *Guide to Rocks and Minerals*. Edited by Martin Prinz, George Harlow, and Joseph Peters. New York: The American Museum of Natural History, 1978.

Muller, H. *Jet*. Butterworths Gem Books. England: Butterworths & Co., 1987.

Mullet, C. "Essai sur La Minéralogie Arabe." *Journal Asiatique,* 6eme série, 11-12, Paris: Imprimerie Imperiale, 1868 .

O'Donoghue, M. *Identifying Man-made Gems*. London: N.A.G. Press, 1983.

O'Donoghue, M. *Quartz*. Butterworths Gem Books. England: Butterworths & Co., 1987.

Philon, H. *Islamic Art*. Athens: Benaki Museum, 1980.

Pliny. *Natural History*, vol. 6, book 37. Translated, with copious notes and illustrations, by J. Bostock and H.T. Riley. London: Henry G. Bohn, Bohn's classical library, 1857.

Rouse, J. D. *Garnet*. Butterworths Gem Books. England: Butterworths & Co., 1986.

Sayegh, S. *Islamic Art*. Beirut: Dar Al Ma'arifa, 1988.

Schumann, W. *Gemstones of the World.* Translated by E. Stern. New York: Sterling Publishing Co., London: N.A.G.Press, Ltd. (in association with Northwood Publications Ltd.), 1977.

Sinkankas, J. *Emerald and other Beryls.* Arizona: Geoscience Press, 1989.

Steingass, F. *Persian-English Dictionary.* Reprint. Beirut: Librairie du Liban, 1975.

Taburiaux, J. *Pearls: Their Origin, Treatment, and Identification.* Translated by D. Ceriog-Hughes. Suffolk: N.A.G. Press, 1985.

Tolansky, S. *The History and Use of Diamond.* London: Methuen & Co., 1962.

Webster, R. *Gems: Their Sources, Descriptions, and Identification.* 4th ed. Revised by R. W. Anderson. England: Butterworths and Co., 1983.

Webster, R. *Gemologists' Compendium.* 6th ed. Revised by E. Allan Jobbins. New York: Van Nostrand Reinhold Company, 1980.

Zamamah, A. Q. "Azhar al Afkar fi Djawahir al Ahdjar," *Majallat al-Madjma'a al 'Ilmi al 'Arabi.* 39, no. 1 (January 1964).

INDEX

NOTE: Arabic words and title of Arabic books in italics. The "al" prefix means "the"; "abu" means "father of," and "ibn" "son of." The inverted apostrophe represents a guttural sound not present in English. In Arabic, diacritical marks are used to distinguish between similarly written letters.

About the Author

Samar Najm Abul Huda was born and raised in Beirut, Lebanon, where she obtained a B.A. degree in Political Science from the Lebanese American University in 1971. She studied gemology while living in Athens, Greece, in the 1980s and is currently researching Arab lapidaries of the eighth to fourteenth centuries. Mrs. Abul Huda has now returned to Lebanon where she lives with her husband and two daughters.